INTERNATIONAL COMMUNICATIONS
Blueprint for Policy

INFORMATION RESEARCH AND RESOURCE REPORTS

VOLUME 4

Editors-in-Chief:

N. SZYPERSKI

and

F. WINKELHAGE

Gesellschaft für Mathematik und Datenverarbeitung mbH Bonn

NORTH-HOLLAND
AMSTERDAM · NEW YORK · OXFORD

INTERNATIONAL COMMUNICATIONS

Blueprint for Policy

KENNETH W. LEESON
East-West Communication Institute
Honolulu
Hawaii
U.S.A.

1984
NORTH-HOLLAND
AMSTERDAM · NEW YORK · OXFORD

ISBN: 0 444 86877 1

Published by:
ELSEVIER SCIENCE PUBLISHERS B.V.
P.O. Box 1991
1000 BZ Amsterdam
The Netherlands

Sole distributors for the U.S.A. and Canada:
ELSEVIER SCIENCE PUBLISHING COMPANY, INC.
52 Vanderbilt Avenue
New York, N.Y. 10017
U.S.A.

Library of Congress Cataloging in Publication Data

Leeson, Kenneth.
International communications.

(Information research and resource reports ; v. 4)
1. Telecommunication policy--United States.
2. Telecommunication--International cooperation.
I. Title. II. Series.
HE7781.L43 1984 384'.068 84-1603
ISBN 0-444-86877-1

PRINTED IN THE NETHERLANDS

INTRODUCTORY REMARKS

The two papers published in this volume of the series "Information Research and Resource Reports" represent related contributions to an ongoing process of discussion and decision making about telecommunication policy in the United States. They are of particular interest because they explicitly deal with the complex interrelationship between various lines of technological development in the field of telecommunications, the cultural, social, economic and security issues involved, and the organizational and informational basis for designing and implementing national and international strategies for telecommunications and information.

Both papers present two main arguments. First, technological, economic and political developments necessitate a coherent, "strategic" view of the field of telecommunications. Second, this strategic view must be supported by organizational and technical measures in order to coordinate and to improve the US posture in the various international bodies where different elements of an international framework for information and telecommunications policy are determined.

From a European perspective, the views and recommendations expressed in these papers may sometimes be alarming in their outspoken determination to shape the future development of international telecommunications according to US economic interests. They make it very clear, however, that the various technical, legal and economic issues have to be regarded as elements in a global political arena where basic values and national or regional options for cultural, social, and political development are at stake, and that they should receive high level political attention – not only in the USA. Thus, it becomes obvious that "de-regulation" of national telecommunication monopolies is not just a question of speedier innovation, improved customer service and lower rates, but foremost a question of organization of national policy making and of the ability to design and to implement telecommunication policies with a view to desired cultural, social and economic developments.

While the paper by Kenneth W. Leeson is an original contribution, the second paper is an excerpt from a report entitled "Long Range Goals in International Telecommunications and Information – An Outline for United States Policy". This report was prepared by the US Senate Committee on Commerce, Science and Transportation in cooperation with the National Telecommunication and Information Administration. In making the selection from this report, it was attempted to avoid duplication, but to include those sections which complement the paper of Kenneth Leeson, and to retain, as much as possible, the line of reasoning of the original document. Thus,

- the second part of chapter two on "Goals, policies, and strategies",
- chapter three on "US participation in the International Telecommunication Union / ITU",
- the chapters on "Trade in equipment and services", "Information", and "National security", and
- the Appendices

of the report on "Long Range Goals ..." are omitted here. The reader is referred to the introductory part of chapter two, which gives a summary of the principle findings and recommendations of these sections. The paper by Kenneth Leeson presents additional, detailed material pertaining to the chapters which are not included in this publication.

Series Editors

CONTENTS

APPENDIX B
Long Range Goals in
INTERNATIONAL TELECOMMUNICATIONS AND INFORMATION
An Outline for United States Policy
(Selected Excerpts)

APPENDIX C
Long Range Goals in
INTERNATIONAL TELECOMMUNICATIONS AND INFORMATION
An Outline for United States Policy

Preface

At the time of writing, my involvement in government with international policy on communications and information had spanned the period from early spring of 1980 to the summer of 1982, just over two years. For most of that time my principal concern was with issues that fell under the catch-all heading "transborder data flow," a field ushered in by computer-to-computer communications that permit the transmission of large amounts of information from one country to another at very high speeds.

By the end of 1980, the center of focus of international discussion on this issue was gradually shifting from concerns with information about individuals -- personal data and the protection of privacy -- to broader questions surrounding the economic, social, and legal aspects of international flows of information of all sorts, personal and nonpersonal, and in particular, commercial information. Two international agreements regarding personal data had been adopted: the "Guidelines on the Protection of Privacy and Transborder Flows of Personal Data" of the Organization for Economic Cooperation and Development (OECD) and the "Convention for the Protection of Individuals with Regard to Automatic Processing of Personal Data" of the Council of Europe. Each of these was the culmination of several years of discussion and negotiation among countries concerned with finding ways of protecting the privacy of their citizens when data identifying them are processed on computers outside national borders.

In the meantime, the mid to late 1970's, governments began giving increased attention to the growing significance of information -- its creation, processing, storage, and transmission -- as a component of economic activity. Statistical studies had been published that affirmed the coming of the "information age," characterized by ever growing proportions of national outputs coming from activities involving the production or dissemination of information.

Information "explosions" and "revolutions" were common hyperbole used to describe the phenomenon.

A cadre of international civil servants in consultation with national representatives were busily trying to establish study plans that would shed light on the impacts of the "new" phenomenon -- impacts on jobs, on competitiveness, on industrial structure, on growth and development. Questions were raised regarding national security, cultural sovereignty, vulnerability of computer and communications systems to natural disaster and crime, and vulnerability of societies attributable to storing strategic information in computers beyond national boundaries.

The more traditional components of the larger picture of communications and information -- telephone, telegraph, broadcast and other mass media -- were in the throes of transformation. Technological advances were rewriting the ground rules on which point-to-point communications had been organized, forcing a reconsideration by individual countries of the viability of the institutional arrangements that had evolved to provide these and a growing variety of new services. "Natural" monopolies were giving ground to competitive forces. In response, regulatory and institutional reform proceeded in different countries at different rates, creating opportunities for new forms of trade and investment among countries in both equipment and services; and for novel ways of closing off markets to competition from abroad.

The World Administrative Radio Conference (WARC) in 1979 initiated what will be a decade of international discussion on issues such as the allocation and use of the electromagnetic spectrum and of parking slots for satellites in geosynchronous orbit. What had been in prior years a technical exercise, with decisions based largely on engineering considerations, was showing signs of becoming a political exercise with decisions based increasingly on political considerations.

In the mass media of word and image, furor erupted over suggestions that journalists be licensed by governments. In response, representatives of the media from a score of nations met at Talloires, France, in 1981 to reaffirm uncompromised support for a free press.

With all of this as backdrop, studies were commissioned by governments and regional organizations; the findings of some emphasized the need for new strategies to bolster capabilities in communication and information-processing activities. Some groups were anxious to begin composing elements of a new international regime of written rules that would regulate information flows. Countries at various levels of development, motivated by desires to reap the advantages that the "information revolution" seemed to promise, began planning strategies to assure their own futures in high-technology industries. There came calls for a "new world information order." A specialized international stratum of aficionados was humming with excitement and activity.

In the United States, concern with more immediate problems began to mount among representatives of industry associations and of individual companies that had either been directly affected by a restriction on their use of foreign telecommunications facilities or were fearful of developments that seemed sure to lead to such restrictions. There was a good deal of coverage in trade journals and specialized publications about potential restrictions on international flows of information. There was trepidation, for example, that privacy protection laws were a veiled form of commercial protectionism and would be used to impede flows of information to advance the economic interests of the countries involved. Professional and trade organizations established committees or working groups to monitor the issues.

There seemed to be a growing feeling that the technological lead in communications and information processing enjoyed by the United States was facing an increasingly serious challenge, a challenge that could lead to gradual though significant economic damage, both to individual companies and to the economy in general. Rumblings that the government was not adequately prepared to respond to the threats posed by policies of foreign governments gathered momentum. They grew into loud cries. There is no "policy" was the gist of the complaint. The Executive Branch was viewed as being so poorly organized on these issues that it was virtually impossible to determine where to find assistance for problems encountered in communications or information flow in foreign dealings.

A good many views on these issues found expression in front of Congress. Hearings were held and legislation aimed at reorganizing the executive branch was introduced. Inspired by such attention, the Executive Branch took steps to improve coordination and to improve performance in meeting the needs of those seeking assistance.

At about the same time, a pronounced reaction developed to the current or anticipated "protectionist" policies of other countries: reciprocity. Though hardly a novel element of United States trade policy, it gained new prominence and a growing field of enthusiasts who sought to clarify or strengthen its role through legislation. This tack was supported by those who saw the need to have in reserve additional clout for responding with reciprocal actions to the degree of access permitted United States suppliers of communications goods and services in foreign markets.

All of the events and developments described thus far, from the allocation of frequency spectrum to trade relations involving "services" enterprises, are related to one another. Yet most will agree that a thorough understanding of the essential elements of this vast field and how they all fit together is lacking. The problems involved in formulating policy are indeed formidable.

Acknowledgements

This paper was prepared at the East-West Communication Institute in Honolulu, Hawaii during the summer of 1982. I gratefully acknowledge its support and the cooperation of the National Telecommunications and Information Administration for granting the necessary leave of absence from my work in Washington. These organizations bear no responsibility for the content of the paper and do not necessarily endorse particular views or opinions expressed within.

While I take full responsibility for the organization of the material and for any views or opinions that may stretch beyond objective observation, I cannot claim credit for many of the underlying ideas. For these I acknowledge a number of contributors, principal among them my colleagues at NTIA, Robert Aldrich, C. Randall Jacobson, and Jane Yurow, and those from other government agencies with whom I have worked on some aspect of the issues over the past two years. Editorial contributions by Helen Shaw and technical assistance by Steven Pomerantz proved invaluable. In addition, ideas were discovered, exchanged, debated, and reformulated with individuals from private industry and through groups such as the Discussion Group on United States Policy on International Information Flow of the Council on Foreign Relations, the Task Force on Information Flow of the United States Chamber of Commerce, the Working Group on Transborder Data Flow of the State Department's Public Advisory Committee on International Investment, Technology and Development, and the Congressional Leadership Group on International Communications of the Georgetown University Center for Strategic and International Studies.

INTERNATIONAL COMMUNICATIONS

Blueprint for Policy

Is it possible to derive from the vast field of activities that comprise "international communications"[1/] a systematic understanding of principal issues and of their significance for social, economic, and political interests? Will such an understanding provide sufficient guidance for the formulation of consistent and comprehensive policy?

The first question can be answered in the affirmative if we are willing to accept as "systematic understanding" a comprehensive listing of individual issues with annotations, grouped in a fashion that will show the relation of each to the others and to the accumulated total. Table 1, to be discussed later, presents one method of accomplishing such an arrangement.

The second question is less tractable. While a systematic understanding of individual issues is likely to provide helpful guidance for making decisions, there are a number of additional conditions that must be met before consistent and comprehensive policy can be achieved.

Policy in International Communications

In this analysis, "policy" will refer to a program that specifies both the objectives to be attained and the means of achieving them. The necessary, if not

---

1/ The phrase "international communications" as used throughout this paper covers issues in telecommunications, information processing, and the mass media; equipment, facilities, services, and information. In search of comprehensive policy it would be counterproductive to begin by embracing or excluding only pieces of the broader picture, though traditional institutional arrangements and the prevailing "division of labor" among those who deal in communication issues seem to encourage such dismemberment.

sufficient, components of such a program for international communications include:

(1) An overview of the state of the world regarding communications and of the forces working to alter that state.

(2) An identification of the individual issues, the components that together comprise the whole of communications activities and, to the extent possible in an era of rapid technological change, an understanding of the relationships among them.

(3) A hierarchy of objectives ranging from a global view of how the overall state of the world in communications should be encouraged to evolve to the desired conditions surrounding individual issues.

(4) A measure of political commitment establishing a "pecking order" of priorities that will serve as the basis for mediating conflicts that arise from incompatible objectives.

(5) An organizational structure of authority and responsibility for decision making to oversee the execution of this program.

(6) Strategies of diplomacy and negotiation for the implementation of decisions taken.

In seeking to devise consistent and comprehensive policy on international communications, governments will have to give some attention to each of these six points. The remainder of this analysis is devoted to a discussion of each in turn.

These clearly are not neatly separable points whose details and component parts can be considered in isolation from one another or resolved in strict sequence.[2/] However, if we wanted to evaluate an international communications policy (or sought to devise an effective one) we would hope in viewing the whole to be able somewhere to identify, at a bare minimum, these six attributes.

2/ Indeed, a good deal of the underlying factors of these points cannot be "resolved" in advance at all but must be dealt with only when the moment to make a decision arrives and in the light of the accompanying circumstances that are undergoing continual change. The "pecking order" of political commitment in item four is an example. A rigid algorithm for resolving conflict among competing objectives cannot easily be specified in advance.

For the discussions that follow, it will help to think of these attributes as falling into one of two groups: those that are prerequisites of policy, roughly points one, two, and three, and those that involve the implementation of policy, points four, five, and six. The distinction is similar to that between thought and action, preparation and execution. The first is an analytical exercise requiring study and planning, the second a political exercise requiring commitment and pursuit.

AN OVERVIEW OF INTERNATIONAL COMMUNICATIONS

Discussions of policy on international communications are liable to strike out in any number of directions. Engineers may ponder the latest advances in transmission speeds or memory capacity and speculate about the probable rate of their introduction; equipment manufacturers may question how the procurement policies of foreign telecommunications authorities will affect their export sales in coming years; strategic planners in multinational corporations will exchange views on the consequences for the location of subsidiaries and of services provided of laws on privacy protection for legal persons; network architects from international financial institutions will consider the implications of having to do without private leased communications lines; government officials of some countries will seek to clarify the effects of international computer-to-computer communications on employment levels or of direct broadcasting by satellite on their national sovereignty; some will raise the issue of the special problems encountered by developing countries trying to enter the age of information; others may warn that future vintages of technology will render insignificant the distinction between mass communications and point-to-point communications, raising difficult questions for freedom of expression.

As these examples illustrate a disucssion of relevant issues may hone in on technological minutiae or wander among broad concerns with changing economic relationships and social conventions. The field covers a wide range of subjects and touches on the activities of all kinds of organizations and of individuals of various backgrounds.

A Simple Abstraction

A helpful step in attempting to devise policy on so broad a spectrum of issues would be to formulate first a simplified abstraction of the underlying forces at work internationally. Thus, the first component of the program outlined above calls for "an overview of the state of the world regarding communications." Here is one rendition:

- Rapid technological change in high-speed computer processing and in communications speeds and capacities has created new opportunities for private enterprise and for individuals; new ways of doing things, new products and services.

- Enterprises and individuals vigorously exploit these opportunities and in so doing invoke changes in their behavior, activities, organizational structures, and the structure and competitiveness of the industries with which they are associated. They do so within individual countries and on an international scale. 3/

- Governments have come to recognize the growing strategic significance of communications and information for their economic, social, and political interests. With such considerations in mind, they do not always approve of the way enterprises and individuals, both foreign and domestic, exploit and adjust to the new technological opportunities. They may take various steps through laws or regulations to promote, discourage or otherwise control the adjustment.

- These actions of government amend the "rules of the game" and affect the available opportunities. Enterprises and individuals

---

3/ In some countries and in some sectors of activities involving communications, governments themselves are the exploiters of new technological opportunities. In the case of postal, telephone and telegraph activities (PTT's) for example, the government is both service provider and regulator and thus plays a dual role.

> adjust accordingly; additional technological opportunities present themselves; there is a reassessment by governments, and so the cycle repeats itself over and over.

This simplification of what is an enormously complex process admittedly buries significant shade, nuance, and exception. Yet it serves the intended purpose. It is this broadly-conceived process, currently underway in the field of communications, that is, technological change, adjustment by private entities, reaction by government, readjustment, etc., from which the international "regime" under which information flows among nations gradually evolves. It is this process and the accompanying evolution of the "regime" that constitutes the backdrop for every decision made regarding policy on international communications. With this as an overall view, the scope of policy on international communications necessarily includes the following five major areas and their components: technological knowledge and its diffusion; physical components, telecommunications and information - processing equipment; the structure of communications facilities and networks; services offered and the terms and conditions of use of facilities and networks; and the treatment of communicated information itself.

The Evolving Regime

The fundamental issue toward which the formulation of policy on international communications is directed, can be stated as follows:

> In light of the changing technologies of communications and information processing and of the growing significance of information in economic activity, how will the "regime" of international and domestic laws, rules, regulations, and practices under which information flows among countries evolve in the coming years?

A regime, of course, exists now and is undergoing continual change. It is the nature and direction of that change that concerns policy makers. In its current form, the regime is not a single, negotiated document or treaty, the product of formal consultations among governments, although there are voices calling for just such global negotiations. At the moment, the regime under which information flows is a diverse mixture of individual practices, laws, rules, regulations, treaties, conventions and agreements pertaining to a vast array of issues. These may have been promulgated solely for domestic considerations by sovereign governments, in the interest of comity by regional or international organizations, or with some particular objective in mind, as in the case of a traditional or novel business strategy or practice followed by private enterprise. All of these together -- and they apply to technology, equipment, facilities, and information itself -- comprise the current regime under which information flows among nations.

What thrusts consideration of this regime into the spotlight now (and helps explain what is behind the forces working to alter the state of the world regarding communications, as noted in the first point of the program outlined above) is an apparent perception on the part of many governments of the growing economic and strategic importance of information.

Although the real significance of the changing role of information has yet to be satisfactorily explained, it is now a widely-held belief that technological advance in communication and information processing has begun to initiate dramatic changes in economies, ranging from working conditions and productivity and employment to rates of development or comparative advantages in trade. When these economic considerations are accompanied by concerns such as national security, cultural diversity, and the vulnerability of computerized society, governments feel compelled to take action, including moves to promote or protect their information and the means of producing and disseminating it.

Real or imagined, the mere perception of the growing significance of information has contributed to an acceleration in the rate of evolution of the regime and hence to the rate of change of the state of the world regarding communications.

COMPONENTS OF THE EVOLVING REGIME

The second point in the program on policy outlined above calls for an identification of individual issues and the relationships among them. Table 1 lays out in summary the three major components of the evolving international regime of information flow and shows how they fit together. The first major component consists of some 35 representative issues selected on the basis of their prominence in international discussion or their significance for international affairs in communications. They are arranged in columns I-V. The second major component, column VI, is a list of the broad areas affected by communications activities, or that serve as motivations for setting communications policies. The third major component, found in the lower portion of the table, shows categories of rule-making that underly the regime.

Categories of Individual Issues

The individual issues in columns I-V are distributed among categories that from left to right represent an idealized progression of the major ingredients of the communications sector:

> (I) issues regarding technological knowledge, the foundation for each category that follows; (II) issues involving physical equipment, the individual building blocks of the communications infrastructure -- computers, switching devices, telephone sets, microwave

Table 1 COMPONENTS OF THE EVOLVING INTERNATIONAL REGIME OF INFORMATION FLOW

Individual Issues Classified According to:

I. State of the Art; Technological Knowledge and its Diffusion	II. Physical Components; Communications and Information-Processing Equipment	III. Structure of Communications Facilities	IV. Services Offered; Terms and Conditions of Use of Facilities	V. Information per se
1. Science policy; R&D activity, production of new technological knowledge on communications	4. Domestic industry structure for producing hardware	7. Domestic industry structure for providing communications services	16. Trade in services	28. Press freedom, restrictions on journalists
2. Indigenous national capacities -knowledge base for producing hardware and facilities; for creating, processing, and disseminating information	5. Merchandise trade in computers and communications equipment -market access	8. Facilities planning	17. Rate structures	29. Privacy protection - natural persons
3. Transferring technology on communications	6. Technical standards	9. Correspondent relationships	18. Liability for economic loss due to system failure	30. Privacy protection - legal persons
		10. International satellite system	19. Use of public networks	31. States as data subjects
		11. Regional satellite systems	20. Direct broadcast by satellite	32. Property rights on information
		12. Allocation of spectrum	21. Value-added networks	33. Valuation and taxation of information
		13. Allocation of satellite orbital slots	22. Interconnection of equipment	34. Surveillance of content
		14. Vulnerability of systems	23. Shared-use networks	35. Remote sensing by satellite
		15. Technical standards	24. Cryptography	
			25. Private leased lines	
			26. Resale of transmission capacity	
			27. Integrated Services digital network (ISDN)	

VI. Effects of Use or Application of Communications and Information Technologies

a. FREEDOM OF EXPRESSION
b. FREE FLOW OF INFORMATION
c. NATIONAL SECURITY
d. BALANCE OF FLOWS AND DEVELOPING COUNTRIES
e. SOCIETAL ADJUSTMENTS TO INFORMATION TECHNOLOGY -HOME AND WORKPLACE
f. VULNERABILITY OF SOCIETIES
g. NATIONAL SOVEREIGNTY
h. CULTURAL DIVERSITY
i. DEVELOPMENT OF INDIGENOUS COMMUNICATIONS INDUSTRY
j. GROWTH AND DEVELOPMENT
k. PRODUCTIVITY AND EMPLOYMENT
l. BALANCE OF TRADE; TERMS OF TRADE
m. FOREIGN DIRECT INVESTMENT

A. Legal Issues -- Choice of law; extraterritorial application of domestic law (Contract arrangement; dispute settlement; liability)

B. Decisions and Practices of: Private Entities; Corporations; Producers; Users, Indigenous and Foreign

C. Actions of: Domestic Lawmakers and Regulators

D. Intergovernmental Understandings, Agreements, Treaties

Kenneth W. Leeson, "International Communications: Blueprint for Policy," East-West Communication Institute, 1983 transmitters, optical fiber cable, etc. This equipment is combined according to technological rules, negotiated technical standards, and institutional conventions to form the communications networks, the facilities up and running. Establishing the structure of these facilities raises the issues listed in column III. These facilities provide communications services to users, and do so under certain terms and conditions, raising the issues listed in column IV. Finally, as a consequence of all these factors working in combination, information flows among countries. Issues concerning the information per se are listed in column V.

With a few exceptions, each of these 35 issues is sufficiently circumscribed to permit the assumption by government of a fairly unambiguous "position" vis-a-vis the issue, though for reasons of strategy ambiguity may be the position opted for. A "position" is a view of the state of affairs one wishes to see surrounding the issue and its subparts, or characteristics one wishes the issue itself to assume. This may include a general specification of the degree of seriousness or gravity with which the issue is viewed or comment on the fact of it being an issue at all or, depending on the nature of the issue, it may include methods of establishing price, terms of availability or distribution, technological content, appropriate source of control, procedures for negotiation, and so on.

At any moment there is an international status quo on each of these issues. Some will tend to change more rapidly than others, will drift in and out of the limelight at different times, and will evoke varying measures of attention and concern among the interests affected.

Effects and Motivations

The second major section of the table is column VI. This contains the principal areas affected by the use or application of communications and information technologies -- the outcomes of the first five columns taken in combination. In one sense, the relationship between the elements of the first five columns and column six is essentially that of cause and effect. Positions taken or practices followed regarding elements on the left will cumulatively shape the attributes of the issues listed in column VI. Conversely, the sought-after attributes of the elements of column VI serve as the motivations for striving to establish a certain configuration of "positions" on the elements of the first five columns. For example, a government may wish to bring about a desired state of affairs regarding its national security which may prompt it to take a certain position or invoke an action regarding, say, cryptography (column IV) or vulnerability of systems (column III) or science policy on communications (column I). Or a private enterprise may have definite plans for the configuration of its overseas subsidiaries (e.g., its foreign direct investment) that will prompt it to take certain views on the conditions of use of public networks (column IV) or on laws regarding privacy protection of legal persons. In one sense, column VI specifies the areas that feel the effects of the combined positions taken by all countries on the elements of the first five columns; in another sense column VI specifies the concerns that motivate governments to act on the elements of the first five columns by setting positions that they may pursue with varying amounts of resolve and ambition.[4/]

4/ The balance struck by any government between passive adjustment and active pursuit will depend on its understanding of the prevailing circumstances surrounding each issue, on its perceptions of what the immediate and longer-term effects of actions regarding the issue are likely to be, the level of concern or significance inputed to the ways the issue develops, and so on. These kinds of considerations, however, are ingredients of strategy about which more will be said later on.

There are, of course, issues beyond the field of communications that affect the elements listed in column VI, some, no doubt, with greater significance than those shown in columns I - V of the table. Corporate tax laws of governments pertaining to foreign enterprises, for example, may have greater weight in decisions by firms regarding direct foreign investment than, say, local "in-country" rate structures for communications services. Similarly, strivings to achieve certain characteristics for the elements of column six may provoke actions by governments on issues well beyond those covered by the scheme of columns I-V. The role of defense expenditures in national security is one example. Table 1 thus provides a view of the relative role of communications issues as factors affecting the concerns of column VI. It might be helpful for the purposes of this analysis to provisionally consider all other factors as going unchanged, though such an unrealistic assumption would have to be dropped by the practicing policy maker.

Strata of Rule-Making

The third major section, in the lower portion of Table 1, shows the institutional arrangements underlying the international regime of information flow. These include (A) legal conventions establishing procedures for contract arrangement and dispute settlement, (B) decisions made and practices followed by private entities, corporations, producers, and users, both indigenous and foreign, (C) actions of domestic law makers and regulators in individual countries throughout the world, and (D) international understandings, agreements and treaties. All of these in combination (and undergoing continual change), through their effects on the 35 issues of the fine categories above them, are the conventions and institutions, formal and informal, that comprise the international "regime" that establishes how information flows among countries.

The first stratum shown, (A) legal issues, is not of a kind strictly comparable with strata (B), (C), and (D). These are the legal issues that arise every day in the normal course of commerce and trade among nations. Perhaps the best way to characterize this set of issues is as those for which one would seek legal counsel, to draw up a contract or to avoid or settle a dispute or conflict that has arisen in some international dealing. Thus, stratum (A) is essentially the practical manifestation of the three classes of practices and rule making of strata (B), (C), and (D) as they exist at any given time.

Relations Among the Parts

The relationships among the three major parts of Table 1, as portrayed by the arrows, represent the "simple abstraction" formulated earlier: technological change, adjustment by private entities, reaction by governments, readjustment and more technological change, etc. Although cause and effect, or action and outcome, work in two directions for any pair of the three parts shown, the predominant flow in the table is circular, in a clockwise direction. The constellation of individual issues of columns I - V affect the issues listed in column six; aspirations for or dissatisfaction with the effects in some of the areas listed here motivate private entities and governments to act to change various of the institutional arrangements of the emerging regime, shown at the bottom of the table. Changes in the regime can either be comparatively modest changes in the rule or laws or practices surounding a single issue, such as an increase in the relative price of private leased lines in a single country, or more comprehensive, such as a major reorganization through legislative means of an entire domestic communications sector. This may affect profoundly the *status quo* of several of the individual issues shown. The outcomes of all such changes influence the issues of column six and further adjustments are sought. Technological advance is the primary instigator of these changes.

The question posed in the first paragraph of this analysis sought an understanding of the significance of the issues now enumerated in Table 1, "for social, economic, and political interests." Even though the table is not organized according to these categories, every issue listed exerts some influence in each of these areas. Noting that fact on the table would not be especially informative. The kind of insight or understanding sought regarding such broad categories of concern will come through thorough study of each of the individual issues involved.[5]

The Appendix contains brief annotations describing each of the approximately 50 individual issues listed in Table 1.

FORMULATING OBJECTIVES

The third component of the program for policy on international communications outlined in the first few pages of the paper calls for "a hierarchy of objectives ranging from a global view of how the overall state of the world in communications should be encouraged to evolve to the desired conditions surounding individual components." This is perhaps the least tractable of the topics to be taken up in the analysis. Along with the statements "should be encouraged to evolve" and "desired conditions" comes the suggestion of a shift

---

5/ "Transborder data flow" is nowhere listed on the table as a separate and distinct issue as it is not. The topics that most often arise in discussions of transborder data flow are scattered throughout the table -- privacy protection, foreign direct investment, valuation and taxation of information, employment and productivity, free flow of information, and so on. In reality, however, the way information flows among countries, which presumably is what transborder data flow means, is the collective consequence of the entire set of issues portrayed here.

in the nature of the analysis from an exercise in objective description to an exercise in political prescription.

The thrust of the discussion here pertains to some of the general problems involved in devising "global views" or general objectives. Consideration of each of the scores of "local" or sub-objectives required for the individual issues of Table 1 and their subparts, while an essential component of the program, would be too ambitious an undertaking for the current exercise and will require a good deal of specialized expertise.

Popular Consensus

The transition from description to prescription would be less difficult to make were there a clearly identifiable "national consensus" on international communications in the sense of broad popular support for a particular global view. Unfortunately, there has been no broad-based national discussion of the issue and thus no such consensus revealed. This is not surprising, however, as individuals without some immediate and conscious interest in the development of communications -- apparently a large majority of the population -- do not concern themselves with this seemingly abstruse issue on so grandiose or general a scale or, if they do, have not effectively expressed the views that they may have.[6/]

6/ This characterization is perhaps unfair. It would seem reasonable to infer, for example, that there is a national concensus regarding such fundamental issues as the sanctity of freedom of expression. "The First Amendment, prohibiting government interference with an individual's right to speak or write freely, provides the foundation for U.S. information policy . . . (and) stands for the principle of open information exchange, as well as providing encouragement to individuals and private organizations for the generation and collection of whatever information is of use to them." Jane H. Yurow, Issues in Information Policy, Special Publication of the National Telecommunications and Information Administration, U.S. Department of Commerce, Washington, D.C., February 1981, p. 1.

Business Views

This lack of popular consensus is accompanied by a demonstrated lack of consensus -- though certainly not of concern -- on the part of the business community in the United States. Of course, diversity of views on any subject from a group as heterogeneous as the business sector is quite natural. Noting a lack of consensus, however, is not to suggest that there has been no discussion or debate on the issue in some quarters, or that there are no strongly held views on how the regime should be encouraged to evolve, or at least certain parts of it. Some corporations that have been able to clearly identify their own interests have indeed taken an active part in working to influence the development of policy. For the moment, however, these are in the minority, and their positions cannot be assumed to necessarily represent the views and interests of others, particularly those of different industries or of firms of different size or facing different kinds of markets. Furthermore, major subsectors of interests may assume opposing positions. To cite a well-worn example, users of communications services and providers of communications services may have quite different aspirations for the nature of the evolving regime and may support rather different global views. Even within each of these groups there can be found inharmonious objectives.[7/]

A further difficulty in trying to draw guidance from the business community in establishing a global view for consistent government policy in this area is a degree of vacillation between wishing a meager government presence in

---

7/ There is one fairly universal business sentiment that may provide some guidance in establishing objectives -- the sentiment that favors less uncertainty over greater uncertainty. Here the overriding emphasis is on the fact of selecting an objective rather than on the precise nature of the objective selected. Affirming a position and sticking with it, as the argument goes, contributes to an atmosphere in which decisions can be made with less uncertainty and risk.

dealings on international communications on the one hand and wishing a very effective and influential presence on the other. Where any given firm or industry association sits in this cycle often follows closely the emergence and resolution of specific problems that may abruptly upset that status quo and which require government assistance for rectification. When the problem ebbs, so may enthusiasm for government attention. Aside from short-term problems, views on government involvement in less immediate, longer-range developments evoke a range of sentiments. There will be more to say on this issue of government role later on. This question of short-term fix versus longer-term developments, however, deserves further elaboration here.

The Short and the Long Term

There seem to be two separate but related strains of international developments in communications. The first takes the form of sudden impositions of new practices or abrupt shifts in policy by other governments that have some immediate impact on the status quo. Such occurrences are usually identified as "barriers" to the use of data-processing or communications facilities and may affect a particular company or group of companies by imposing some unexpected and perhaps sustained hardship. The second strain is the more gradual evolution of the communications policies of other countries, individually and collectively, which may have no visible or immediate impacts and which may not affect directly any particular interests in an obvious way, but which in the long term will exert their toll as they come to establish the international conventions under which information flows. Both of these strains are part of the evolving regime under consideration in this analysis.

Efforts of government agencies in tending to the former kinds of problems can be viewed as analogous to the efforts of most any private firm in its quest to meet short-term goals, such as next quarter's profits. Government attention to the second strain of issues can be viewed as analogous to a private firm's strategic planning, that is, its preparation now for those short-term pay-off periods that will inevitably arrive sometime in the intermediate or far-off future. As seems the tendency with private corporations in general (or at least according to popular notion) the government is in this matter of communications far more adept at dealing with the short term than in preparing for the long term. As an example, when a company finds itself in some way harmed in another country by a clearly discernible or even subtle restrictive practice on communications, deemed unfair under the terms of existing international treaties or conventions or prevailing perceptions of acceptable practice, the company can seek assistance through the Departments of State or Commerce or the United States Trade Representative; the circumstances of the case will be clarified and rectification of the problem sought. There are established procedures available for treating these kinds of problems.

In recent times, this short-term strain of developments in international communications -- identifying individual problems on an as-they-arise basis and then trying to fix them -- has served as the basis for most of the constructive debate on the adequacy of government policy.[8/] That is, the debate revealed specific problems regarding the government's performance and bore practical suggestions for solutions to those problems. These came most often in the form of schemes for reorganization or suggestions about consolidation of authority and responsibility. Unfortunately, there has not been a comparably constructive debate with regard to questions of performance of the government on longer-

8/ This "debate" took the form of periodic advisory meetings between representatives of companies and government officials, speeches delivered and discussion undertaken at conferences of various kinds, articles and editorials in trade publications, and congressional hearings.

term strategic considerations in international communications. On this aspect of policy, what little discussion there may have been was effective only in specifying a general problem, essentially pointing out the lack of strategic preparation in the area. Few practical suggestions for alleviating the problem have been forthcoming beyond, perhaps, vague statements, often delivered with an air of resolve and gravity, that something needs to be done here.

The distinction between the short-term and long-term strains of the issue as characterized here, becomes apparent from a reading of "International Information Flow: Forging a New Framework,"[9/] one of the most concise and thorough compendiums available of the issues involved in international communications policy. It was prepared on the basis of a good deal of study and analysis, interviews with government officials and representatives of private companies, and of congressional hearings that elicited views and concerns from a wide spectrum of industry and academia.

For the short-term issues, such as existing or new barriers of various kinds to international flows of information, the report provides careful detail on the sources and nature of the problem, then offers some concrete recommendations on how the executive branch of the government might improve its performance in responding to requests for assistance from affected corporations.

On longer-term strategic considerations it is far less effective and of marginal use to the task laid out in this section of identifying global objectives. It states that "The United States must remedy its problems of preparation and organization. The executive agencies must plan, establish priorities, develop policy, coordinate efforts -- in brief, initiate an effective scheme for

---

9/ Thirty-Second Report by the Committee on Government Operations, United States House of Representatives, 96th Congress, 2nd Session (U.S. Government Printing Office, Washington, D.C.: December 1980). Hereafter referred to as "Forging a New Framework."

representing American interests."10/ When the crux of a recommendation is to "plan," "establish," "develop," and "coordinate," it becomes essentially a solution in search of a solution. Thus, in spite of all the expert testimony and analysis behind the report, there emerges no clear guidance on what the broad objectives of an international communications policy would be.11/

The absence of a broad popular consensus on objectives; incompatible interests among private groups that have significant stakes in how the international regime of communications develops in coming years; the lack of clearly articulated objectives beyond suggestions on organization provided by recent congressional activity; mercurial consternation regarding the implications for the balance between public and private roles of a more consistent and orderly government presence in policy on international communications; and the perennial revelation that any action taken in the name of communications stands a chance of being thwarted either by the pressure of competing national goals or through the absence of capitulation on the part of other countries -- all of these together should provide ample discouragement for pursuing any further an exercise intended to identify global objectives. Perhaps it would be appropriate at this point to conclude that for all practical purposes it is not possible to articulate an acceptable hierarchy of objectives or perhaps not politically or socially desirable to do so. It may be preferable to choose instead to rely on the current procedures, born out of an institutionally imposed division of labor in

---

10/ "Forging a New Framework," p. 56.

11/ The major thrust of the hearings and of work that produced this report were directed toward questions of organization and not toward questions of the substance of policy, so it is not surprising that the major outcome has been recommendations for reorganization. Nevertheless, the fact that there emerged from all of that effort and the tiers of experts assembled few hints as to what the strategic objectives of policy might be provides us with a good feel for where we stand with regard to the state of the art on the issue.

which scores of decisions are made monthly in various enclaves of government on a more or less ad hoc basis with no clear view of an overall direction for communications policy.

The situation is perhaps not as bleak as this portrayal suggests. There are two important considerations that have been glossed over thus far that may provide useful guidance in the search for objectives. The first is, as just noted, that decisions on a wide array of factors that affect the issues in this evolving regime are made everyday, and although there may currently be no single "expressed" global view shared among the decision makers, each presumably is acting on the basis of some notion or conception (perhaps not explicitly stated, but generally shared) of the ultimate implications of the decisions. A search among existing documents and an analysis of actions taken could well reveal whence decision makers are finding their inspiration. By inference we might be able to construct a picture of underlying objectives. A second factor that could well provide some lessons in global objectives for international affairs is the domestic communications and information "regime" in the United States, perhaps the healthiest, most efficient, and innovative in the world, providing a large variety of reliable services under conditions of unimpeded flows of information.

On this first point, some progress has already been made. On the second, adopting domestic objectives as international objectives, there exist a few significant caveats, but the notion is essentially a sound one.

Objectives by Inference

Compiling for all fifty issues of Table 1 a complete inventory of the legislated directives, regulations, administrative memoranda, and position papers that have served as the basis for past decisions would be an onerous yet essential step in working toward coordinated policy. All that could be gleaned regarding the nature and efficacy of underlying objectives, either stated or "inferred" from decisions made, as well as insights regarding consistency would constitute an essential element of the repertoire of the members of the "organizational structure of authority and decision making (that would) oversee the execution of this program," (item five of the program on policy). Here it will suffice merely to point out the need for such a painstaking exercise and not assume the burden of actually carrying it out. It will be useful, however, to review briefly a few of the existing sources that contain broad objectives or general guidance for the formulation of policy on international communications.

The document that comes closest to laying out a global view of the way the international regime should be encouraged to evolve is an informal paper prepared by William Salmon of the Department of State.[12/] Of particular interest is the fact that this document was prepared by means of the inferrential method described above. The author and contributors analyzed current and past decisions and activities of the government on international communications and surmised from these analyses seven basic objectives. These were then distributed "to help stimulate discussion among American constituencies of (international aspects of communications and information)."[13/] They are also examples of what may be authentic specimens of the type sought for this component of the program on policy. All seven are reproduced here. The Salmon paper states that:

> Broad U.S. objectives respecting the international communications and information issues reflect a diversity of interacting domestic and overseas interests. The world where these objectives must be pursued is characterized by widely divergent

---

12/ "International Aspects of Communications and Information," William C. Salmon, Interim Coordinator for International Communications Policy, Office of the Undersecretary of State for Security Assistance, Science, and Technology, February 12, 1982. Hereafter referred to as the Salmon paper. Sections reproduced here with permission of the author.

13/ Salmon paper, cover page.

philosophies, cultures, capabilities, and political and economic systems. Moreover, technologies continue to evolve. We are, therefore, involved in a process that can be expected to continue for many years to come.

The seven basic objectives below, reflecting clusters of interests, are identified as goals toward which government actions could be directed. They are inter-related. Their order does not indicate relative priority.

(1) <u>To enlarge acceptance of the principle of free international flow of information and ideas, including applicability of this principle to newly emerging communications and information technologies.</u>

From the U. S. standpoint, the free flow principle embodies the objectives of the First Amendment to the Constitution. Ongoing changes in communications and information technologies have great potential for fostering the growth of more open societies worldwide. Through greater openness, mutual understanding among nations can be improved and economic growth of all nations benefitted. More effective

management of resources, more beneficial trade policies, more effective responses to global problems, and better maintenance of international peace and security are achievable through such openness. Therefore, although this principle is of special historical and philosophical significance to the U. S., it is relevant to the interests of all nations

(2) <u>To expand the economic benefits of communications and information technologies by broadening opportunities for competition and investment.</u>

Communications and information technologies, services, and products are now supplanting in actual economic importance many more traditional goods and comprise an increasingly significant element of the U. S. economic strength. The U. S. must seek -- and should expect --opportunities for its industries to compete fairly. The U. S. cannot ignore the growing evidence of protectionist practices in various forms in many countries, a trend which, if not checked, could limit the potential values of the new technologies to all countries.

(3) <u>To ensure the flexibility and continuity of communications and information required to maintain national defense and international peace and security.</u>

Effective communications and information resources are of fundamental importance to strong U. S. military capabilities, for deterrence and defense, arms control and peacekeeping efforts, and contribute greatly to international peace and security. Equally important roles are played in maintaining security by linking friends and allies and by serving as routes for resolution of frictions and conflicts, particularly during periods of dispute and crisis with adversaries.

(4) <u>To ensure equitable access for users of the radio frequency spectrum and orbital positions.</u>

Modern telecommunications technologies play a central role in all countries. Access to the various bands of the radio frequency spectrum and to orbital slots is vital to U. S. economic and security interests; such access is important equally to other nations. The potential of new technologies to support more efficient use of frequencies and orbital positions can assist in ensuring equitable access for all countries.

(5) <u>To enlarge the communications and information capabilities of developing countries.</u>

National economic development, regional political stability and a healthy global economy require effective communications and information capabilities. Developing countries can gain significant economic and social advantages for their internal development, as well as a more effective voice in the responsible conduct of world affairs, through improved communications. Enhancement of commercial and broader economic relationships depends upon effective communications and information exchange relations. Existing U. S. development policies stressing basic human needs must recognize to a greater extent the need for improving communications and information capabilities of developing countries. Greater involvement of the U. S. private sector is also needed.

(6) <u>To stimulate continuing advances in communications and information technologies.</u>

Maintaining U. S. leadership to meet foreign competition effectively requires increasing attention to -- and creation of opportunities for -- innovation. Government actions have a profound effect on the private sector environment for such advancement. The market place should fairly test the acceptance of new products and services; government regulations should not stifle individual initiative.

(7) <u>To improve the basis for developing and implementing policy.</u>

Improving the basis for policy development and implementation requires continuing and increasingly effective relationships between the Executive Branch and the Congress; among the various Executive Departments and Agencies and the Federal Communications Commission; and between the government and the private sector. More effective discussions among the United States and other countries concerning these questions are also essential.

Whether these constitute an adequate global view to guide decision making on issues from each of the columns of Table 1, or whether additions or modifications are required, however subtle or pervasive, will be revealed through further discussion and, most reliably, through application, should a comprehensive program under discussion here come to pass. According to their author, the typical comment on these seven points made by reviewers is that they merely state the obvious. This would tend to support the suggestion raised earlier that policy is currently formulated by informal yet "obvious" notions in the minds of decision makers. In any event, bringing the obvious forward from the back of our minds and laying it out clearly for review seems a step in the right direction.

There are other examples of documents and activities that hit close to the mark by providing glimpses of broad motivations that underly positions on particular issues. The Federal Communications Commission has a regulatory role regarding the authorization and use of international communications facilities and has promulgated objectives to guide policy in this area[14/] and has taken a series of initiatives to "benefit the consumer through increased competition for international record services."[15/] An official report of the Department of State on international communications policy states that "Since the end of World War II, the United States and its democratic allies have strongly supported the free flow of information as expressed by Article 19 of the

---

14/ For example, "Inquiry into the Policies to Be Followed in the Authorization of Common Carrier Facilities to Meet North Atlantic Telecommunications Needs During the 1985-1995 Period." Notice of Inquiry Federal Communications Commission Reports, Volume 73, 2nd Series (U. S. Government Printing Office, Washington, D. C.; September 21, 1979), p. 193.

15/ Competition and Deregulation in International Telecommunications, an Analysis of Fifteen FCC Actions and Their Combined Effects; prepared for the National Telecommunications and Information Administration (U. S. Department of Commerce by Mar Tech Strategies, Inc. July 1981).

Universal Declaration of Human rights, which asserts the right of everyone 'to seek, receive and impart information and ideas through any media and regardless of frontiers.'"[16/] The list can easily be expanded. There are a good number of sources of ideas for establishing the broad views and objectives that will guide the development of policy in coming years. Pulling them to the fore and achieving consensus is another matter.

Extending Domestic Objectives

There are attributes of the domestic "regime" under which information flows in the United States that would seem desirable to adopt as part of a global view for the international regime. Among these are strict prohibitions against restrictions on free expression or on flows of information, and a general preference for reliance on market forces whenever practicable to determine kinds of services provided, quality, and price. It is probably fair to say that these general attributes do indeed constitute important bases for positions currently taken by the United States in international deliberations on issues in communications and in commercial and trade relations in general.

A policy that would seek to simply extend domestic attributes to the international regime would be, for obvious reasons, unrealistic. That is, other than through persuasion by example (or by entreaty), domestic attributes cannot be unilaterally hoist upon other sovereign nations or blocs of nations not of a disposition to embrace them of their own accord. This does, however, prevent such attributes from serving a useful function in establishing the foundation for decisions made and for positions taken on international issues.

16/ "International Communications Policy," Section 601 of Reports Submitted to Congress Pursuant to the Foreign Relations Authorization Act. Fiscal Year 1979 (Public Law 95-426), Senate Committee on Foreign Relations, House Committee on Foreign Affairs, 96th Congress, 1st Session (U. S. Government Printing Office, Washington, D. C.; July 1979).

Some of the pitfalls of attempting to simply extend principles of the domestic regime to the international regime have been noted in two congressional reports, with both comments motivated by concern on this issue in the area of regulated communications facilities. One of the reports points out that "... a major distinguishing feature of overseas communications is the presence of foreign correspondents and the 'mixing together' of this country's telecommunications, trade and other objectives and regulations with those of foreign nations. Within this framework, domestic communications goals cannot and do not predominate. When conflicts with foreign policy arise, the most basic attempts to promote traditional public interest objectives may fall by the wayside. In particular, priorities of American trade, defense, and diplomatic policy often take precedence over provision of 'efficient service at minimum cost.'"[17/]

This raises the issue of constraints, but does not render useless the objective of "efficient service at minimum cost." It merely recognizes that such an objective is worked toward in an atmosphere of competition with other, often incompatible, objectives; as in all other political endeavors, constraints must be reckoned with, but the fundamental usefulness of the original objective as a view of the desired state to bring about remains largely uncompromised by such constraints. Futhermore, the constraints cited in this passage -- objectives regarding trade, defense, and diplomatic policy -- are, as might be argued on the basis of Table 1, not at all separate from issues in international communications.

17/ Telecommunications in Transition: The Status of Competition in the Telecommunications Industry, A Report by the Majority Staff of the Subcommittee on Telecommunications, Consumer Protection, and Finance of the Committee on Energy and Commerce, U. S. House of Representatives, November 3, 1981, (Washington, D. C.: U. S. Government Printing Office), p. 147.

Indeed, such a broad-based view may imply the need for a reassessment of "priorities."

The other report is one cited earlier and lays down some very clear views on one aspect of this question. "The Federal Communications Commission cannot simply extend its domestic decisions to international markets. The tendency . . . to treat international questions as an extension of domestic ones must be avoided."[18/] To help minimize these tendencies the report goes so far as to advocate, "the full separation of staff support in international matters, including tariff decisions and rule makings, from the personnel and resources devoted to domestic matters."[19/]

With these admonitions in mind, it is important to preserve the distinction between the unrealistic matter of extending domestic policies to the international arena on the one hand, and drawing heavily from domestic practices to provide guidance on global objectives in international discussion and negotiation on the other.

Global View for the Practicing Advocate

In light of the considerations on establishing objectives reviewed in the previous sections, it might again be appropriate to ask whether a "global view" is really a prerequisite for the formulation of comprehensive policy and whether it is even possible to devise such a view. In the abstract, talk of "global view" sounds logical and responsible, but, as just demonstrated, in practice the difficulties encountered in devising such a view are formidable.

18/ "Forging a New Framework," p. 49.

19/ "International Information Flows: Forging a New Framework," p. 49.

Let us proceed by reaffirming that a global view is an important component of comprehensive policy on the argument that with or without a global view, decisions on all of the issues of Table 1 will be made in the coming years; it will be more favorable than less to have them made, whenever practicable, on the basis of some common, overall view of how the regime should be encouraged to evolve.

Devising a global view will require experimentation. That is, since analyzing the issues involved has not resulted in the emergence of a clear notion of a global view or what its components would be, then we proceed by suggesting functional global views. Then, through a process of debate and field testing in real-world situations, the views can be accepted, rejected, corrected, or augmented as required. This, of course, is not a wholely novel procedure. In some respects it describes the current method of disjointed policy formulation, the prominent difference being that here we are seeking a common view for all issues and not a list of potentially incompatible views for the different parts of the scheme.

The assumption will be that someone has been assigned the responsibility for implementing a program on international communications policy covering all of the issues of Table 1. The primary function of this fictitious person is to "advocate" positions (or to refrain from advocating positions if it is not impolitic to do so) on any question that may arise (or for long-term preparations that have to be made) on the issues compiled in columns I through V of the table. This person is characterized as an "advocate" because it is unlikely that he or she would have authority to make final decisions on all of the questions that might be raised from so broad a range of issues.[20]

20/ There would, for example, be little or no authority over decisions made by independent regulatory bodies such as the Federal Communications Commission, though positions can certainly be advocated for such decisions.

In working toward devising a global view to assist this policy advocate in discharging responsibilities, it would be helpful to lay out some of the attributes that a broad view should possess. There are at least seven such attributes.

Since the breadth of responsibility is so wide,

(1) the global objective should provide a basis for making consistent and compatible decisions over a broad range of related issues.

Since the evolution of the international regime of information flow is only one among a number of related issues of strategic importance such as foreign policy, defense, economic and social welfare,

(2) the global objective should not compete in any significant or consistent way with other important national objectives or be so rigid as to preclude the possibility under certain conditions of capitulation in the face of conflict.

The evolution of the regime is a gradual process that will proceed indefinitely through time, hence,

(3) the global objective should be sustainable through time and not subject to abrupt withdrawal or significant revision with changes in elected administrations.

The dividing line between international issues and domestic issues is thin at best in communications and, for many purposes, nonexistent. Decisions made in the domestic market will have impacts on the interests of other countries and decisions made by others will be felt in various ways in the domestic arena. In some instances, as is the case with international facilities, the decisions and their effects are taken jointly, thus,

(4) the global objective should be consistent with objectives established for domestic activity and should permit accommodation whenever feasible of the objectives of other countries. (This does not, of course, preclude the taking of positions on individual issues that are in direct conflict with the objectives of other countries. Resolution of such conflicts should be sought through negotiation and a process of give and get.)

As a state of general uncertainty with regard to the course of decision making is perhaps more debilitating of the actions of individual and commercial enterprise than is the consequence of any particular decision made,

> (5) the global objective should reveal a clear general direction for decision making that will help reduce the degree of uncertainty facing entities whose interests are directly affected by the evolution of the international regime.

As this policy advocate will, in any event, be in the unenviable position of advocating decisions on issues that will have direct effects on interests from all sectors of the economy,

> (6) the global objective should be neutral or "blind" with regard to its effects on particular commercial interests (though it would be unreasonable to expect such neutrality with regard to individual decisions). Achieving such neutrality may require that the objective address itself not to specific outcomes for particular issues, but to the conditions under which activities take place and events unfold.

Finally, since the policy advocate must establish and support positions,

> (7) the global objective should be functionally useful, providing some basis for choosing among alternative courses of action and making decisions.

This is a substantial list of desirable attributes, but should not appear too threatening, especially as the tack suggested here to develop global objectives permits experimentation with alternative formulations. It will thus be a trial-and-error process, one whose feasibility will be confirmed or denied over a period of time. The nature of such an iterative process renders it impossible to go much further here with the experiments. Before moving on to other subjects in this plan for policy, however, it might be helpful to posit an example of a global objective. The example provided here reflects some of the characteristics claimed for this evolving international "regime" throughout the analysis.

The following observations borrow heavily from the formulations layed out earlier in the paper:

(1) Timely, accurate, accessible information whether distributed point-to-point or via mass media is the "final product" of the communications sector. 21/ Equipment and facilities, that is, computers, satellites, transponders, cables, microwave towers, networks, video display terminals and so on are valued ultimately for what they can do to facilitate some step in producing, processing, storing, transferring or disseminating information for use.

(2) The terms and conditions under which information is produced and disseminated internationally -- the amount and type available, its speed of movement, quality, accessibility and cost are determined essentially by three related factors:

(a) the state of the art -- in equipment, facilities and information; the latest technologies and earlier vintages.

(b) the social and economic institutions and practices through which these technologies are applied.

(c) the laws and rules promulgated by governments on institutions and practices in order to regulate this application of state-of-the-art technology.

(3) Relegating provisionally to a position of secondary concern all of the activities and interests that serve as intermediate inputs to the production and dissemination of information, that is, not having immediate concern for their interest or condition as factors separate from their role in accommodating the flow of information, then an overall objective for the policy advocate can be established. This advocate will have responsibility for overseeing the development of the terms and conditions under which information flows internationally, with a view toward promoting the widest variety of sources of information, the least expensive and most efficient means of processing and transferring the information, with the fewest possible impediments to its flow, . . . or some similar formulation. 22/

21/ This is the final output of the communications sector and, with the exception of what goes to final consumption, input at some intermediate stage in the production of almost every other good or service available.

22/ This or any other such "global objective" is not necessarily a state of affairs whose precise attributes can be specified in great detail so that we can periodically check to determine if at some particular date in the future it has come to pass. It is more accurately characterized as a comparative criterion, relative rather than absolute, an enumeration of desirable and attainable condition toward which the "regime" should be encouraged to develop (or be pushed).

The first and third observations provide guidance on the overall objectives of policy in international communications. The second describes the basic international processes underway. With these as guidance, this policy advocate sets about reviewing all of the issues of Table 1 and their subparts, establishing or reaffirming positions on each and then planning strategies to support these positions.

The advocate will, of course, at times be cooperating with and at times competing with other interests. It is, after all, the responsibility of other "advocates" to promote the well-being of what are for them issues of first priority (e.g., trade balances, employment stability in equipment manufacturing, foreign investment, security, international relations, and interests of all kinds associated with each of the fifty or so issues listed in Table 1). With regard to a global objective in international communications policy, however, these are intermediate factors that contribute in some way to the production and dissemination of information.

COMMITMENT AND PRIORITIES

The potential for conflict with other interests in the pursuit of the objectives laid out for the policy advocate in international communications raises the issue of the fourth necessary component of the original program for policy: "a measure of political commitment establishing a 'pecking order' of priorities that will serve as the basis for mediating conflicts that arise from incompatible objectives." This issue requires little elaboration.

Conflict will arise whenever the objectives for the regime of international information flow are incompatible with other national objectives. As it is virtually certain that no objective in any area will have unchallenged precedence over all other considerations, it would not be possible in practice to establish a strict plan for resolving conflict in advance of an actual problem. It would, however, be essential to the effectiveness of the performance of this policy advocate to have a generally unambiguous notion of the order of priorities. Developing this notion is largely beyond the province of the advocate. It comes largely from without, determined by deep-seated political philosophies, by prevailing interpretations of those philosophies, by mood or sentiments, and by the effectiveness of representatives of particular interests to influence public opinion and the views and behavior of policy makers. The issue boils down to the level of "cost" a nation stands ready to impose on intermediate concerns and activities that may conflict with the ultimate issue of international information flow; or, alternatively, of the "cost" it stands ready to impose on international information flow when conflicting objectives seem of overriding importance.

STRUCTURE AND AUTHORITY

Following close behind the issue of political commitment to objectives regarding the substance of policy is the issue of commitment to the execution of policy, as expressed in the fifth component of the program: the need for "an organizational structure of authority and responsibility for decision making to oversee the execution of this program." Presumably, the imaginary "policy advocate" created in the previous pages would fit into the structure somewhere, but with what resources and what authority is an issue that remains unresolved.

On this question of carrying out policy, there are three issues that require attention:

(1) the proper role of government in influencing the course of international communications,

(2) design of the organizational structure and the placement of authority and responsibility, and

(3) efficiency and comprehensiveness of the effort in relation to the amount of resources devoted to it.

The most significant among these three is undoubtedly the first. Its resolution establishes the underlying authority and responsibility of government and hence the boundaries for aspirations regarding the second and third issues. In fact, the question of the proper role of government underlies all components of the program on policy laid out in this analysis. It is a subject tied closely to the establishment of objectives as discussed in the previous section and requires only a few additional observations here.

There is a good amount of material already available on the issue of design of organizational structure and placement of authority and responsibility.[23/]

The third issue, efficiency and comprehensiveness of effort, is perhaps the least developed of the three and will receive the majority of attention here.

Proper Role of Government

An underlying assumption of this exposition has been that direct and active involvement in influencing the course of the evolving regime of international information flow is the legitimate province of the Federal government. This is an assumption subject to challenge. Unfortunately, there is currently no unambiguous guidance or mandate, either in law or in practice, that would

---

23/ See, in particular, "International Information Flow: Forging a New Framework," which provides fifteen recommendations on government organization.

satisfactorily settle the issue of the proper role of government in this area.[24/]

The nature of international communications itself prompts questions on the government's role. Is the development of the international regime of information flow an issue of foreign policy that necessitates a strong and organized government presence to respond to the policies of other governments? Is it simply the international version of domestic issues concerning the treatment of information for which strict prohibitions are placed by the First Amendment on the power of government? This would imply the propriety of very limited government involvement. Is the evolution of the regime a commercial issue in which the traditional limitations on direct government involvement in (or interference with) the domestic marketplace should be adopted as guidance for a government role in the international arena? Is the evolution of the regime essentially a trade issue for which the government assumes an active role as negotiator of agreements with other governments and as an intermediary in settling disputes? The problem is that all of these considerations apply to international communications. These circumstances contribute in government both to organizational confusion in assigning responsibility and to a hesitancy to select decisively among paths that involve more or less government involvement.

A more effective government role need not imply a larger government role, either in the sense of intrusion into the domain of private enterprise or in the sense of cost. A view underlying this analysis is that government is already involved in efforts of some kind in most of the issues that have been raised here;

24/ Guidance on the role of government for some parts of the field is more clearly laid out than for others. The regulatory responsibilities of the Federal Communications Commission in international telecommunications, for example, are established by statute (the Communications Act of 1934 and the Communications Satellite Act of 1962). Still, there are differences of interpretation of the breadth and nature of the mandate such legislation actually provides and they cover only certain aspects of the far broader questions being addressed here.

the primary concern is with improving the effectiveness of the role it already has and not necessarily with expanding that role.

Organizational Structure

The array of issues involved in international communications have strong connections with foreign policy, defense policy, regulatory policy, economic and monetary policy, trade policy, employment policy, and domestic communications and information policy. Currently there are cabinet level or other major departmental groups set up with primary responsibility for each. Where, then, should responsibility be placed for international communications policy? Should responsibility be centralized or distributed among a number of agencies? Should it be assigned to an existing niche or a newly created one? Should the issues receive attention at the highest levels or is it sufficient to handle them at intermediate staff levels? Will the erosion of traditional institutional boundaries in the economy and society as a whole brought on by technological change in communications imply the need for some pervasive revision of organizational boundaries in government?

The task of devising a satisfactory answer to the question of where responsibility should be discharged or with what kind of authority, will not be taken up here. It will suffice to proceed with this notion of a "policy advocate" as someone or some group assigned responsibility for carrying out the program on policy being developed here.

Applying Technology: An Information Network

Earlier, the policy advocate was assigned the responsibility of reviewing each of the more than 50 issues of Table 1 to establish objectives and positions consistent with the global views. A brief review of the annotations of those issues contained in the Appendix will confirm that the list of 50 really only skims the surface. On closer scrutiny the number of individual issues grows many times over; subparts and related questions are uncovered at every turn.

The assignment given to the advocate is formidable. The job will require that a vast amount of information be collected and kept up-to-date. There are half a dozen executive departments concerned with some piece of this array of issues involving on modest approximation several dozen separate offices or other departmental subdivisions and hundreds of individuals; among them there are a variety of views about who has responsibility for what; there are now in existence any number of interagency committees and task forces meeting periodically on a broad range of these subjects to discharge responsibility for some piece of the larger picture; various kinds of advisory committees meet to provide advice from private industry; there are delegations leaving Washington weekly bound for one international meeting or another to present United States "positions" on relevant issues; there are everyday major or minor shifts in the policies or positions on these issues of individual countries around the world; there are several international organizations each with an agenda and calendar of activities; Congress holds hearings on bills or proposes legislation that somehow affect the rules of the road in policy on international communications.

How will the advocate effectively discharge the responsibilities assigned? One solution lies in the application of the very technologies of communications and information that raise the issues treated in this paper.

One of the principal determinants of the performance of the Federal government in its efforts to develop comprehensive policy on international communications is the availability of information -- well organized, timely, accurate, detailed and easily accessible -- on the long list of issues and activities that comprise the area. Efforts to improve coordination will be frustrated unless all agencies involved can be kept informed about the substance of issues, developments throughout the world, the activities of other governments, positions and objectives established on individual issues, and the strategies designed to attain those objectives.

Thus, the question is how to remain informed on a vast array of issues and how to ensure that government agnecies with involvement in some piece of the picture have the information they need in order to discharge their responsibilities. This could be accomplished by establishing a computerized information network.

Configuration and Content

The precise configuration of such a network need not be specified in detail at this point, but its salient characteristics can be. A diagram of the kind of system being proposed is provided in Figure 1. Responsibility for the network would rest with the policy advocate who would oversee the development and maintenance of a computerized data base containing all of the necessary "files" of information. Among these files would be:

(1) Activities of the executive branch, including:

- o Diagrams or descriptions of where primary responsibility and authority rest for decision making on each of the component parts of the area.
- o A list of interagency committees and task forces and their functions and responsibilities; a calendar of meetings and their agendas.
- o A list of delegations bound for bilateral or multilateral meetings; a calendar of the meetings and their agendas.

(2) Profiles of the major issues of policy, along the lines of those listed in Table 1, containing for each, information on:

- o Established positions of the United States
- o Specific instructions for delegations scheduled to depart for meetings
- o International forums where the issue is discussed
- o Domestic policies on the issue
- o Economic implications or interests surrounding the issue
- o Political and social implications; implications for national security
- o Positions assumed or actions taken by other countries
- o Existing or imminent problems -- specific cases
- o Proposed recourse and strategies devised
- o Coordinating body in the executive branch
- o Relevant legislation; congressional activity
- o Calendar of activities; meetings with private sector, agency or interagency meetings, meetings of international organizations.

(3) Profiles of the communications and information policies of other countries containing, on a country-by-country basis, information on:

- o General attitudes and directions
- o Outlines of policy-making structures in government
- o Characteristics of their domestic market -- policies on competition
- o Positions taken in international meetings
- o Bilateral contacts with the United States; dates, issues, outcomes

INFORMATION NETWORK
FOR POLICY ON
INTERNATIONAL COMMUNICATIONS

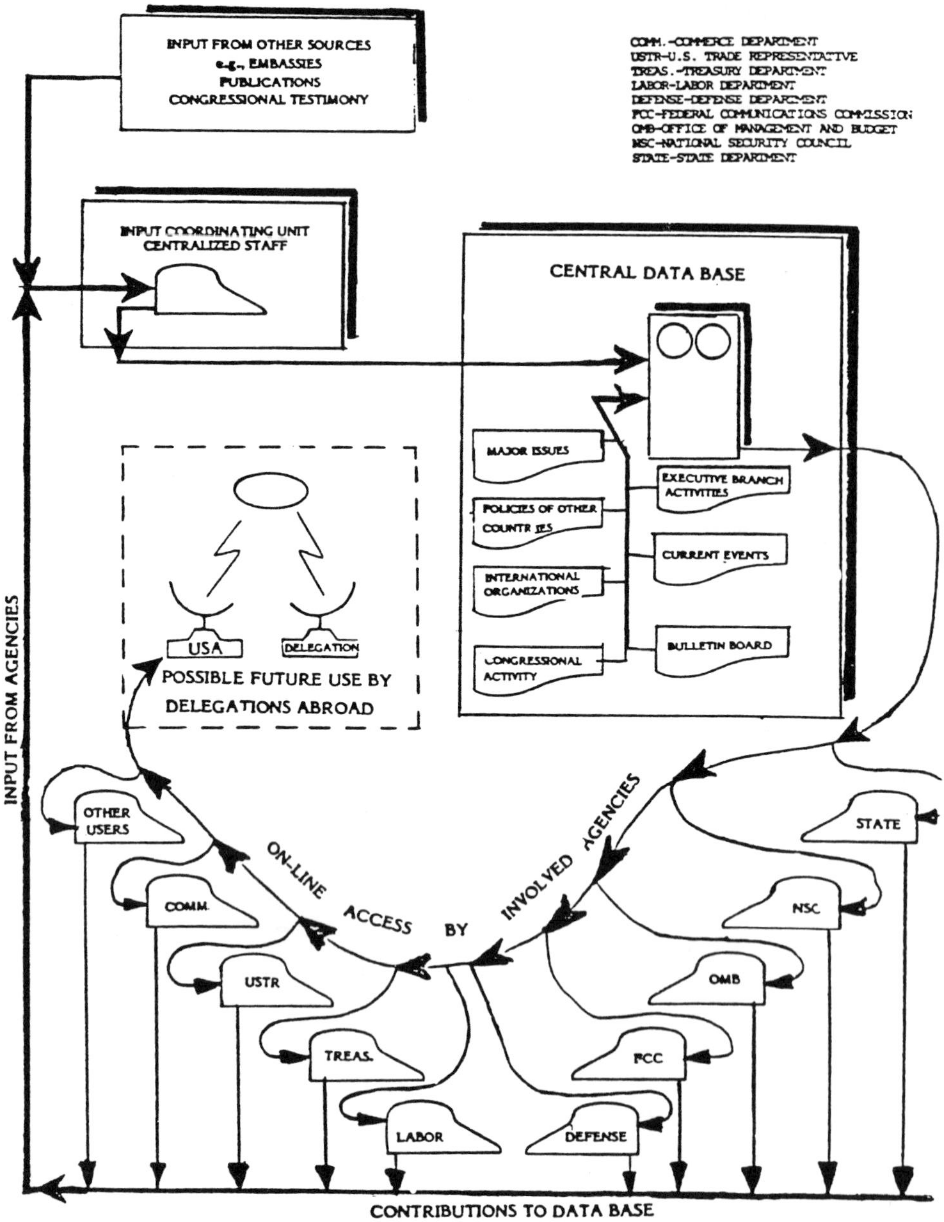

- o Enumeration of their laws, rules and regulations regarding communications and information.

(4) Profiles of international organizations containing information on:

- o History and current status
- o Membership
- o Kinds of issues handled; agendas
- o International status or primary function (forum for discussion and exchange of views, promulgator of non-binding recommendations, or of legally binding treaties)
- o Calendar of upcoming activities
- o Executive branch agencies involved
- o Positions of the United States.

(5) Laws, in place and pending, that affect the issues and determine positions to be taken; calendar of congressional hearings.

(6) Current events, and late-breaking news obtained from the press and trade publications and from cable traffic from United States embassies around the world.

(7) Bulletin board, and mail box facility for communications among individuals, sending messages, making general announcements, clearing cables going to embassies or delegations, clearing executive positions for congressional testimony.

Operating Characteristics

The data base would be centrally located, with each agency having direct access to it by on-line terminals. Responsibility for upkeep of the data base -- adding new information and purging the obsolete -- would rest with the policy advocate and there would have to be some procedure by which each agency or user could submit information forinclusion on an ongoing basis.

Establishing the technical characteristics and type of information the system would actually hold would be the responsibility of its users. Consideration would have to be given to other issues, such as whether the system will carry classified information and whether it should be designed for access from terminals outside of the country for use by delegations needing immediate information during discussion or negotiation.

Cost and Benefit

One of the most sobering considerations in a proposal such as this is the anticipated cost. The project has to be viewed, however, just as any other investment that requires an initial outlay of resources. If the stream of benefits derived over the life of the system outweighs its costs, then, depending on the length of time it takes to reach such a break-even point, the availability of funds, and the relative significance of competing projects that could also make use of the funds, the project may be worth pursuing. The benefits in this case include not only reductions in inefficiencies and duplicative effort, but less quantifiable returns, such as its contribution to developing an international climate conducive to achieving a broad range of social and economic objectives.

This project could be initiated in a piecemeal fashion, gradually building up from a modest system with limited capabilities to one containing increasingly sophisticated attributes, with assessments along the way as to the relationship between costs and benefits. Deciding on whether to proceed and in what fashion is, again, primarily a question of political commitment.

STRATEGY

The sixth and final component of the program on international communications policy calls for "strategies of diplomacy and negotiation for the implementation of decisions taken." Strategy directs action in support of goals and objectives.

Specific statements or recommendations on strategy cannot be made until objectives are specified, political commitment declared, and resources and organizational structure determined. In the absence of these prerequisites, the procedure will be simply to raise the kinds of questions that would have to be considered in devising a strategy and to make a few observations about the circumstances facing the strategist.

Questions Posed

Is a "from-the-top-down" approach preferable to a "from-the-bottom-up" approach? That is, should global negotiations be undertaken among all countries with a view to specifying the general terms and conditions of international information flow, codifying them, and then having individual countries and private entities adjust their behavior accordingly; or should an atomistic approach be supported where the regime is left to develop piecemeal, with hundreds of decisions and agreements made among countries, bilaterally and in the appropriate international forums, on individual topics and according to the particular circumstances surrounding affected issues?

Is preemptive negotiation preferable to reactive negotiation? That is, even on individual issues, should discussion and agreement be carried out in anticipation of problems, or would this be too stultifying and costly and the identification of potential problems too difficult? Would it be more effective to wait until tangible problems have arisen and then engage in discussions or negotiations with the countries and entities involved to try to solve the problem? Are there characteristics peculiar to certain issues that render them compatible with one or the other postures -- preemptive or reactive -- and should they be so divided?[25/]

Has the time come for negotiations on newly emerging issues of information flow? That is, should discussions on emerging issues (those not now formally treated in international forums for setting standards of various kinds) be sought now with an eye to setting standards, or would such a move be premature?

Decisions made on the basis of greater understanding of the forces involved are preferable to decisions made on the basis of less understanding. It will be important, therefore, to proceed with discussion and negotiation from a position of an adequate, if not solid, basis of understanding of the changes being made by technological advances. Has such understanding been achieved or should study and analysis, discussion and debate be encouraged, rather than a rush to rule-making?

Should the general posture assumed and demonstrated in international discussions be active or passive? Should delegations go to international meetings prepared with initiatives, suggestions, and proposals (strongly established positions on the issues of Table 1 which would be, presumably, well coordinated through the efforts of the policy advocate) or would low visibility and less initiative be preferable? What tactical advantages regarding the unfolding of events are there inherent in being the proposer of an idea, rather than having to react to or argue against the proposals of others? Should an effort be made to participate actively in studies, commissions, and international deliberations of

---

25/ A similar point was touched upon in the discussion of objectives; short-term fix versus longer-term strategic preparation.

all kinds on the argument that influence in directing the course of such deliberations can more effectively be exerted; or should participation be avoided on the argument that the results of such efforts may ultimately have to be disavowed because of their conclusions or findings? This would be an awkward turn of events for any country that has participated all along in an effort.

Would it be more desirable to maintain the current institutional separations among major blocks of the whole field of international communications -- telecommunications, trade, the mass media, computers -- and act on them separately; or are these separations becoming obsolete and meaningless or even misleading?

What is the most effective way to deal with coalitions; regional (e.g., European Communities), ideological (e.g., "Third World"), or institutional (e.g., Conference on European Posts and Telegraph, CEPT)?

To what extent would interests be better served by pursuing bilateral discussions and negotiations rather than working through multilateral organizations? For which particular issues is one to be preferred over the other?

Which international organizations should be selected as appropriate vehicles for treatment of each of the various issues?

Should consideration be given to particular circumstances and needs of various regions of the world? For example, are the issues the same and do they require the same strategies for, say, the Pacific region and the Atlantic region; for developing countries in Southeast Asia and those in Africa or South America, or do they require separate consideration?

What is the most fruitful approach to the issues with regard to countries at different stages of economic development? In which forums should the issues be raised with these countries?

In the pursuit of objectives, how should influence be exerted, and what are the most efficacious methods in each circumstance (by setting an example, through promise of reciprocal treatment, threat of retaliation, actual retaliation)?

What is the proper timing for each issue? When is an issue ripe for action and when is it right to refrain from action?

In deciding on the issues for which negotiation and formal agreement should be sought, what will be the trade-off between their positive effects in reducing uncertainty and risk and, thus, presumably improving the climate for doing business, and their negative effects of imposing too much structure on a technologically dynamic area, stifling innovation and reducing entrepreneurial opportunity?

Answers to these questions will have to be sought in light of some of the following observations.

Is the Phenomenon New?

Information has always served as a significant component of economic activity and social intercourse. Is information and the means of transferring it today fundamentally different from the way it was ten, twenty, or thirty years ago? This is a valid question, but does not have to be answered directly. It is sufficient to observe that the perception on the part of some governments is that the significance of information is changing and that this has led to the development of novel policies, laws and other actions that affect interests beyond their borders. Whether we are witnessing an information "revolution," "evolution," or "devolution" may be the subject of idle speculation; that visibly significant steps are currently being taken in reaction to the phenomenon is cause for attention and, in some instances, concern.

Contributions to Understanding

The nature and effectiveness of the program on policy will bear a strong relationship to the amount of resources available for planning, analysis, preparation, and implementation. Facing the realities of budgetary constraints, the strategist will have to devise methods, in cooperation with industry, of tapping the private sector for assistance. This would constitute an important source of knowledge, understanding, and up-to-date information. The level of active participation and support by private interests (subject, of course, to legal constraints) will be an indication of the perceived significance of the effort to the interests that stand most to be affected. The design and success of the program will depend heavily on input from private interests.

National Differences and International Comity

One of the objectives of international discussion and negotiation is to promote a climate in which individual nations can effectively participate. The extent to which any given country is prepared or suited to participate in international activities will depend, in part, on the degree to which the international regime replicates its domestic regime. It is not surprising to observe, therefore, that in negotiations countries attempt to infuse their own familiar habits, conventions, and practices into the international sphere. The greater the national differences in economic conventions (e.g., rules of private enterprise), legal structures (e.g., common law, statute law), and social traditions, the more difficult is likely to be the effort at striking compromise.

International negotiations undertaken to devise acceptable international "standards" for the protection of personal privacy illustrate this proclivity to promote national models. Most of the activity in this area has been among industralized countries for which there exist two different legal traditions:

those of the common law countries (e.g., the United States, Canada, Great Britain) and those of the civil law countries (e.g., continental European countries, Scandanavian countries). These have produced different approaches to the protection of privacy. The essential difference for this illustration is that the former tradition relies more heavily on the evolution of laws and practices through historical cases, judicial interpretation, and contracts, while the latter relies more heavily on the enactment of statutes that formally codify laws, often in anticipation of problems -- codes of conduct. In the case of the United States, privacy law comes from a variety of sources including the Constitution, the common law and statutes and regulations at the federal and state levels and gives different treatment to different kinds of personal records depending on their purpose and the industry or activity involved. By contrast, in most of the European countries that have privacy laws, the approach has been to enact a single "omnibus" statute giving uniform treatment to records held for any purpose either by government or in private data banks and regardless of purpose or industry. Commissioners are appointed to oversee the application of the statute. This often requires registration of data banks with the central authority.

An international problem arises when personal information flows among countries that take these different approaches to privacy protection. Whose standards should apply? Will the standard of protection offered in one country be met in others? If not, are the purposes of the laws compromised? Should this justify restrictions on flows of personal data to keep them in-country where protection is enforceable? There have been two international efforts to answer these questions, one undertaken by the Organization for Economic Cooperation and Development and the other by the Council of Europe.26/ The former resulted in voluntary guidelines which serve to permit endorsement of certain minimum standards of privacy protection and which recognize that different countries may take different legal approaches to achieving the standards. The latter effort produced a legally binding treaty, one to which only countries with "omnibus" privacy laws can accede.27/

In future discussions and negotiations on other aspects of the evolving regime, the problem of differing legal, economic, and social traditions among countries will have to be dealt with time and again. The outcomes of negotiations may be a move toward "harmonization" of laws, with countries adjusting their domestic practices to international requirements, or "accommodation," with different national approaches maintained and tolerated. There is, of course, a third possibility: no satisfactory resolution and continuing discord.

Random Shocks and the Fallacy of Composition

One of the desirable attributes outlined in the section on objectives was that the global view be sustainable through time and not subject to abrupt withdrawal or significant revision with changes in elected administrations. In reality, the strategist is going to have to contend with decisions made at high political levels for reasons of foreign policy or national security that may

---

26/ See the Appendix for annotation V.29, "Privacy Protection, Natural Persons."

27/ It is interesting that the United Kingdom, a "common law" country has decided to embrace the "civil law" approach to privacy and has signed, but has not yet ratified, the Council of Europe Convention. There are plans to enact the kind of omnibus privacy law that would be required to abide by the terms of the treaty. See "Data Protection, the Government's Proposals for Legislation," British Home Office, London, April 1982.

introduce unexpected inconsistencies into the plan. The likelihood of developments of this kind bring to the fore the problem of a fallacy of composition in the symmetrical scheme of Table 1: the whole may be something quite different from the sum of its parts.

Technology and Existing Institutional Boundaries

A strategy that would seek to base international agreements and practices regarding the treatment of information on existing institutional arrangements may tend to unnaturally perpetuate industrial structures that technological advances would otherwise cause to change. This problem is faced in the domestic setting as well. Frequently, regulatory schemes based on conventional technologies and traditional notions of which commercial activities are the proper domain of which industries, inhibit the natural evolution of industrial structure. The lines that divide banking services, retail merchandising, communications services, and data processing, for example, are maintained more by regulatory fiat than by technological or economic necessity. As in the case of establishing technical standards, there are trade-offs here. Agreeing on the rules for and boundaries surrounding the activities of industries establishes predictable procedures and thus reduces uncertainty, but it also inhibits the natural development of innovative institutional arrangements that would otherwise be brought about by the technological change.

Technology as Final Arbiter

It has been argued that technology will foil attempts by governments to exert effective control over international flows of information. There will be so many communications channels (cable, microwave, and satellite circuits) reaching into each country and tying it to the rest of the world. Digital transmission methods will homogenize all messages (voice, record, data, video), producing an indifferentiable stream of binary signals transmitted through packet-switched networks that send bits and pieces of each message over different routes for reassembly at the destination. Cryptographic techniques will be improved. The miniaturization of components, such as silicon chips, will make it possible to store large amounts of information on minute media for convenient transport. Because of these developments, as the argument goes, it will be infeasible for governments to enforce laws that may tend to restrict, intentionally or unintentionally, communications of any kind among countries. There may be something to this argument. However, there are good reasons for not relying solely on technology to be the final arbiter of international relations. Using technology to circumvent the laws of a country may merely encourage the development of technologies of surveillance. These, though costly and disruptive, may go a long way in closing technological loopholes. Secondly, enforcement of laws may be accomplished through unannounced audits, where violations of laws may be uncovered irrespective of the technologies used in everyday operations. Thirdly, for corporations with foreign-based subsidiaries, complying with the laws of a host country is simply a matter of good business practice. If violations of laws on communications are discovered, the penalties could jeopardize the overall standing of the corporation, in the country of violation and elsewhere.

Negotiating as Number One

When dealing from a position of technological or economic advantage, it is wise to approach the question of negotiation with caution. If, relative to the other parties to the negotiation, a country has most to give, and already has most of what it is liable to get, then striking a satisfactory bargain may require a good deal of careful forethought. In fact, the safest option may be to abstain from negotiations altogether. This situation of advantage does not characterize the United States for all issues dealt with here, but it does for many. Where it does not, and where the negotiation is less risky, strict limitations on the breadth of coverage should be agreed upon in advance, in order to avoid having to make inappropriate or non-germane concessions.

Many negotiations in this field -- correspondent relationships, for example -- are not of this zero-sum variety, where what one negotiating partner gives up the other acquires. The motivations and criteria for entering into negotiations of this latter sort may be quite different.

Defensive Postures versus Participatory Cooperation

Protectionist policies are motivated, in some instances, by a desire to build domestic capabilities in emerging industries. Restrictive postures are likely to be assumed by countries that consider themselves underdeveloped in the technologies of information processing and communications. Thus, the greater number of countries that have developed their capacities and who thereby perceive it to be in their own best interest to support a liberal, nonrestrictive international régime, the more free and open the regime is likely to be. Assisting countries in their attempts to enter the information age is an important consideration of long-term strategy.

Strategies of Others: Domestic or International?

The motivations of many countries for formulating plans and strategies on communications and information grow primarily from the desire to develop their own domestic capabilities and to reduce dependency on outsiders. From the point of view of the United States, however, which has traditionally had the dominant role as world supplier of the goods and services in this sector, such developments in domestic policies of other governments constitute an international issue. Thus, with a few notable exceptions, other countries have not yet devised international strategies per se, although their domestic policies do constitute an international challenge to the United States. This situation will gradually change as countries become more competitive and are in a position to seek foreign markets for their own goods and services in communications and information.

Competition versus Cooperation

Any strategy will have to contend with two general factors of opposing natures: competition and cooperation. Although developing an efficient worldwide communications system requires a great deal of cooperation among countries in terms of sharing knowledge, establishing standards, and reaching agreements on rules and behavior, there is a large subset of issues, usually economic or market oriented, in which countries are in direct competition with one another. Questions that must ultimately be considered by individual countries include the extent to which cooperative postures can harm competitive capabilities (and how much can be sacrificed) and, conversely, how much the cooperative requirements of a well-functioning global system will be compromised by competitive tactics.

Entreaties for Efficiency

Finally, U.S. strategists have been forced to recognize that in an area such as communications and information processing, in which a "national capacity" is widely viewed as being of strategic importance to a country's economic well-being and security, attempts to dissuade governments from taking restrictive measures based on arguments of "optimization" of global resources may not be very persuasive. Arguments in support of an open international system on the grounds that all will benefit by exploiting comparative advantages, specialization, and an international division of labor, may not be very persuasive.

APPENDIX A

COMPONENTS OF THE EVOLVING REGIME: ANNOTATIONS

Table 1 lists the principal issues comprising international communications and shows their positions in the emerging regime as characterized in this analysis. This appendix provides annotations of the issues and gives brief synopses of the factors that pertain to the role of each in the overall scheme. In some instances, references are provided to published materials that have served as sources of information or ideas. In other cases, when the issue may be very broad in scope or lacking a well-developed history, the annotations are somewhat less specific. Each is accompanied by a Roman and Arabic numeral to indicate its location in the column of Table 1.

I. State of the Art; Technological Knowledge and its Diffusion

The three issues under this heading -- science policy, indigenous national capacities, and transfer of technology -- constitute the knowledge base from which the technologies and know-how required to build the communications and computer systems emanate. Each of these three issues covers a rather wide range of subtopics, and raises fundamental questions about policy. Efforts to adjust policy in these areas are major undertakings, and the effects of changes in policy are rarely immediately noticable, becoming apparent instead over long periods of time.

I.1. Science Policy; Research and Development Activity -- New technological knowledge on communications and information processing springs from research and development activities, and these can be supported or encouraged by private industry and government through a number of practices and laws. While the United States currently enjoys an edge in basic technological knowledge in communications, there are disturbing signs of erosion of that advantage. Falling behind in "know-how" and its application would have serious implications for long-term competitiveness of the entire sector, and would be felt in other activities that rely heavily on communications activities.

The following questions require attention in devising comprehensive communications policy:

(1) Are the incentives provided in the marketplace for expenditures on research and development adequate, or should government be given greater responsibility here, either by increasing funding for such activities directly, or by underwriting a larger amount of the risk of investment through tax incentives or other means?

(2) Do patent, copyright, and trade-secret laws provide adequate incentive for invention of new information products?

(3) Would it be advisable to liberalize antitrust laws to permit more collaboration by firms on research and development?

(4) What policies would encourage domestic firms to find novel applications of information technology, and to produce commercially competitive products and services?

(5) Is adequate attention being given to education and training to ensure an educated, innovative workforce, and to provide a steady stream of new ideas and their successful commercial exploitation?

I.2. Indigenous National Capacities -- It is widely recognized that the technologies of communications and information processing are vital ingredients of the efficient operation of a wide range of industrial activities and of the maintenance of national security. For these reasons, a number of governments have taken steps to ensure the existence of an adequate indigenous knowledge base for producing hardware and facilities, and for creating, processing, and disseminating information.

National scientific and technological capacity is a critical factor in a State's or a firm's competitive position, according to a report by the Organization for Economic Cooperation and Development. National capacity is defined as "the ability to innovate or promptly incorporate external innovations (into) product ranges or production systems."28/ The concept extends beyond intellectual ability to invent; it includes the industrial capacity to innovate.

The implications of a solid knowledge base in information technologies for a country's international competitiveness are obvious, and have served as the motivation for intervention by governments seeking to ensure the development of indigenous capacities. The long-term implications of other government policies to build up their own indigenous capacities (particularly through restrictive measures on imports or on access to data bases outside the home country) require attention by the United States. Efforts to minimize restrictive practices with the argument of maximizing global economic efficiency may have little effect in an area such as this, in which the strategic significance of developing a viable indigenous capacity may outweigh the perceived benefits of open trade, international specialization, and the division of labor.

I.3. Transferring Technology on Communications -- The transfer of technologies from one country to another has become an important aspect of policies aimed at developing national capacities. Technology transfer (the export of "know-how" by one country; the import of "know-how" by another) can be accomplished in a variety of ways, including:

- exchanges through technical and scientific conferences, and through publications;
- exchanges through the placement of students in the universities and research institutes of other countries;
- acquisitions by firms in one country of a share of or controlling interest in firms in another country, including the technology embodied in their production techniques and products;

28/ Impact of Multinational Enterprises on National Scientific and Technical Capacities: Computer and Data Processing Industry. Ad Hoc Policy Group on Multinational Enterprises, Directorate for Science, Technology and Industry, OECD, Paris, 1977, p. 150.

- establishment of subsidiaries in other countries by a parent corporation that sends with them knowledge of technique for producing a product or service. This knowledge is presumably acquired by the work force of the host country;

- joint ventures undertaken by indigenous firms with foreign firms that may include, for example, provision for the transfer of know-how in exchange for knowledge of or access to markets;

- licensing of patents by firms in one country to firms in other countries; and

- reverse engineering, in which a product of a firm of one country embodying a technique is dissected by a firm in another to determine how it works.

Policies adopted on the methods of transfer may vary according to: (1) the place held by the technology on the spectrum that ranges from basic scientific understanding to direct applicability in productive (and hence competitive) endeavors; (2) difficulties encountered in entering product markets directly which makes licensing of patents a more practical alternative; (3) national security concerns about the "dual uses" of some technology for both non-military and military applications;[29/] (4) perceptions of the likelihood of success of actions taken to selectively control the transfer of know-how; and (5) the desire to share knowledge with countries at different stages in the development of their national capacities.

There are both cooperative and competitive aspects to transferring the technologies of communications and information processing. It is in the interest of all countries involved to share technological knowledge for the purpose of assuring a smoothly running and integrated global infrastructure; yet the incentives for creating and the means of disseminating much of the required technology is subject to competitive forces and market strategies.

29/ See, for example, the annotation IV.9 on cryptography.

II. Physical Components; Communications and Information-Processing Equipment

The components, or building blocks, of communication networks and information-processing systems play a significant part in determining the nature of the regime under which information flows. These physical components are important in their own right, as well, because the production and sale of such equipment in both the home and foreign market are important factors in economic growth, employment, and trade.

II.1. Domestic Industry Structure for Producing Hardware -- A country's capacity to produce and disseminate information is dependent in large part upon the availability of computer and communications equipment. The health of the industry that produces this equipment will influence the pace at which information technology is applied to other economic activities.

Among the issues that require review here are the industrial policies of governments that affect the structure and viability of the domestic sector that produces hardware.[30/] These include regulations, antitrust exemptions, channeling of research and development funds, direct government involvement, tax treatment, export financing, etc.

If this sector is considered strategic to the country's future, are special policies designed to encourage its growth and development appropriate? What should such policies be? What should policy makers seek to accomplish with regard to the nature of trade relations -- the terms of importing and exporting such goods?

30/ Structure refers to number of firms in the industry, their size, extent of vertical integration, skill requirements, and the availability of a trained work force. Viability refers to the industry's long-run profitability, its ability to offer products at reasonable cost to buyers, its ability to remain on the frontier of technology, and its capacity to compete, both with foreign suppliers in the home market and with the indigenous suppliers in foreign markets.

II.2. Merchandise Trade in Computers and Communications Equipment-Market Access -- In 1980 the United States exported $25 billion in communications merchandise -- physical products used in the production, processing, storage or transmission of information. Examples of these products include telephone apparatus and switching devices, computers, radio and television receiving and transmitting apparatus, engineering, laboratory, and research equipment, etc.[31/] High-technology products, in which the United States has a strong, but perhaps eroding, competitive edge over other countries, account for approximately 70 per cent of total exports of such equipment.

For many of these products, export markets absorb a significant proportion of total industry output. In 1980, for example, exports of electronic computing equipment alone totaled over $4 billion, nearly 25 per cent of the output of that industry in the United States. A significant portion of producers of equipment rely on access to foreign markets for sales, access that may be threatened by the trade practices and information policies of other governments.

It is important to consider whether the competitive advantage enjoyed by equipment manufacturers is slipping away. Aspects of this problem include:

- The adequacy of domestic industrial policies and the practices of firms regarding research and investment. Will such policies assure the maintenance of a competitive industry?
- Identification of the tariffs, quotas, and non-tariff barriers that create obstacles for equipment manufacturers attempting to export their products.
- The terms of access to the home market offered to other countries.

31/ Kenneth Leeson, United States Trade in the Merchandise of Information Industries, Vol. I, in Trade Issues in Telecommunications and Information, National Telecommunications and Information Administration Publications, April 1981, and updated working files.

- o Identification of the markets of tomorrow and the role the government should play in promoting access to those markets.
- o The extent to which efforts of other countries to build up their own equipment-producing capabilities, in some cases through protectionist measures, affect trade and the health of the domestic industry.
- o The appropriateness of reciprocal trade arrangements as a response to the policies of other governments.

II.3. Technical Standards -- Setting standards is an attempt to establish common practices of all kinds that will enable components of the global communications system, regardless of location or country of origin, to connect and operate. Technical standards on physical components of the communication infrastructure refer essentially to the engineering characteristics of equipment that permit physical coupling and thus the establishment of information systems or communications networks.[32/]

Technical standards that vary from country to country create barriers to trade. Exporters will not find buyers in a foreign market if their equipment cannot be integrated into the importer's communications systems. Thus, the use of technical standards to inhibit the importation of foreign-produced equipment is a subject of significant concern in international trade relations. The General Agreement on Tariffs and Trade (GATT) now contains a "Standards Code,"[33/] obliging signatories to refrain from imposing unnecessary obstacles to trade through the use of standards and testing requirements. Uniform international standards for telecommunications and information equipment, reciprocal acceptance of performance tests, and openness of certification procedures will

---

32/ The issue of technical standards on components is closely related to the issue of standards and protocols on facilities, for which an annotation is provided under III.9, Technical Standards.

33/ The Agreement on Technical Barriers to Trade (Relating to Product Standards), General Agreeement on Tariffs and Trade, MTN/NTM/W/192/Rev.5., March 29, 1979.

all serve to reduce barriers to trade.

Title IV of the Trade Agreements Act of 1979[34/] outlines the obligations of Federal and state agencies and private persons in the United States regarding technical standards. These obligations include avoiding "unnecessary obstacles" to foreign trade, ensuring that imported products are treated on the same basis as domestic products, using international standards, if appropriate, as a basis for developing new standards, developing standards on the basis of performance rather than design criteria, and allowing foreign suppliers access to certification systems on the same basis as that extended to domestic suppliers.

The perennial questions to be asked on the issue of standards include the following:

- o Whose standards should be adopted?
- o What are the benefits of establishing technical standards that permit the widest possible interconnection of equipment?
- o What are the drawbacks of institutionalizing rigid practices that may stifle innovation and impede the introduction of new technology?

III. Structure of Communications Facilities

The International Telecommunication Union (ITU) is the principal international forum for establishing agreements on the structure of telecommunications facilities and procedures for their use. As part of the ITU structure, the International Telegraph and Telephone Consultative Committee (CCITT) studies and makes recommendations regarding technical characteristics of telecommunications systems, their operations, and pricing. The International Radio Consultative Committee (CCIR) has similar responsibilities for radio

---

34/ House of Representatives Document 96-153, Part II, "Trade Agreements Act of 1979 - Statements of Administrative Action," U.S. Government Printing Office, June 19, 1979, pp. 211-233.

communications. ITU's International Frequency Registration Board (IFRB) accommodates the orderly allocation of the frequency spectrum among users. Numerous references to ITU recommendations in the annotations in this section and the next, particularly to those promulgated by the CCITT, demonstrate the significance of these organizations in determining characteristics of the emerging regime of information flow.

III.1. Domestic Industry Structure for Providing Communications Services

The entire international communications infrastructure can be viewed as having three major parts: (1) the in-country structure of the home country, (2) the in-country structures of foreign countries, and (3) the international structures of cables and satellites that tie them all together. For each part, there are very different conditions or circumstances regarding control and influence under which policy on communications is formulated. The home country has virtually complete control over decisions regarding the structure and use of its own communications sector. It has essentially no direct control over decisions regarding the in-country structure and use of the communications sector of other nations, and has only partial control through cooperative arrangements over the structure and use of the international facilities that link countries together.

From the standpoint of any given country, the collection of domestic communications structures of all other countries combined is a prominent component of the international communications regime confronting it. The actual significance of one country's domestic structure for another country's international interests will, of course, depend upon the extent of commercial, cultural, and political ties between the two.

The industry structure for communications services in the United States differs in fundamental ways from structures found in most other countries.

Telecommunications equipment and services are provided by private enterprises, with the provision of some services (largely common carriage telecommunications and broadcast) subject to government regulation. The growing field of computer processing and "enhanced services" that combine both computer processing and telecommunications go unregulated. The government does not itself provide services or equipment. In most other countries, however, services and equipment are provided directly by (or under the control of) government entities, usually monopolistic in nature and generically referred to as PTTs (for postal, telegraph, and telephone authority). The differences between these two kinds of industry structures are significant, particularly with regard to the degree of competition permitted in (and the sources of control over) the provision of goods and services.

Among the major frustrations in formulating international policy is the difficulty or impracticality of extending beyond United States borders some of the more attractive attributes of domestic policies on communications. These attributes include deregulation and competition, interconnection of equipment with the networks, and the freedom with which information is permitted to flow. Whenever other sovereign countries revise their domestic policies toward the practices followed by the United States, the international regime is evolving favorably. Thus, an important consideration in formulating international policy and in devising strategies is the extent to which the United States can influence such developments through example, persuasive argument, or promises (or threats) of reciprocal treatment.

III.2. Facilities Planning -- International communication links -- undersea cables and satellite circuits -- that connect the United States with other national networks are owned jointly by United States carriers and foreign telecommunications administrations. Each owns half of each link in operation, extending to the imaginary midpoint for undersea cable and from the ground to the satellite for satellite circuits. Establishing international telecommunications facilities is thus a cooperative undertaking involving entities of two or more nations -- private, regulated firms in the case of the United States, and government authorities in most other countries.

Satellite planning for international communications (how many, where positioned, when retired, how used, etc.) is undertaken through INTELSAT (International Telecommunications Satellite Organization), an organization jointly owned and administered by 106 nations. Undersea cable systems are planned for the United States and its overseas counterparts by the American Telephone and Telegraph Company in consultation with United States international record carriers (e.g., Western Union International, RCA Global Communications, Inc., ITT World Communications, Inc., TRT Telecommunications, Inc., FTC Communications, Inc.) and the relevant foreign telecommunications authorities.

By law, the construction and operation of new facilities by United States companies require approval by the Federal Communications Commission (FCC). Technically, this gives the FCC final authority regarding plans negotiated between United States carriers and their foreign counterparts. In reality, however, it is considered politically unfeasible for the FCC to disapprove facilities that have been approved through the international consultations described.

The North Atlantic Consultative Process began in 1974. Participants include representatives of the United States Government, international carriers, and representatives of Canada and of European countries belonging to the European Conference on Posts and Telegraph (CEPT). This consultative process, which is an effort to coordinate planning for cable systems and satellite systems, and to provide a forum for the exchange of information and views, is carried out through informal conferences rather than through formal negotiations. A similar effort for facilities planning in the Pacific region was initiated on a limited scale in 1981.

The effectiveness of the consultative process in resolving conflicts in the planning of facilities has met with mixed reviews. Satellite planning still proceeds essentially independent of the consultative process and there is some question about the authority and proper role of the government in this area.

An important question to consider is whether a wide enough range of alternative facilities is being considered in the facilities planning process, in terms of cost-effectiveness, successful commercial relationships, and greater national security.

III.3. Correspondent Relationships -- The establishment of international telecommunications links requires the cooperation and consent of at least two entities, or correspondents -- the domestic and the foreign. As institutional deregulation plus technological and economic factors in the United States promote the entry of new competitors among the providers of international telecommunications services, it will become necessary to find acceptable methods of reaching "interconnect" agreements between the new entrants and foreign correspondents. This may pose a challenge, given the documented reluctance of foreign PTTs to deal directly with all but the few well-established United States carriers.

III.4. International Satellite System -- INTELSAT, the International Telecommunications Satellite Organization, a consortium of 106 countries, owns, maintains, and operates the global satellite system that provides public international telecommunications services. By most accounts, the system that has developed has been a clear success. As of March 31, 1981, INTELSAT had 13 satellites in synchronous orbit, and carried the majority of intercontinental communications traffic -- including telephone, data transmission and television -- on more than 20,000 circuits. The ground segment of the system consisted of 204 communications antennas in 135 countries, territories, and possessions.[35/]

INTELSAT is owned by member countries with each holding an investment share based on the extent of its use of the system. Revenues are derived from utilization charges and are distributed to members in proportion to their share as amortization of their investment and as compensation for use of capital.

The United States, through its official representative to INTELSAT, the Communications Satellite Corporation (Comsat) has been a staunch supporter of the development of the global satellite system.[36/]

As in other areas, technological opportunities and competitive forces are disturbing these institutional arrangements that had been established to promote development of the international satellite system. The use of non-INTELSAT satellites for regional international transmission and the setting up of non-INTELSAT direct broadcast satellite systems for international business use may be forcing a reconsideration of policies on the international satellite system. Issues to be considered here include whether the legislation that created Comsat in 1962 is in need of revision because of changing technology and a changing competitive situation, and the impact the establishment of competing regional systems will have on the structure and viability of the existing global system.

35/ INTELSAT, 1981 Annual Report, p. 8.

36/ Unlike most members of INTELSAT, the United States is not represented by the government but by Comsat, a privately held corporation established by an Act of Congress in 1962 to be the United States participant in the development of this worldwide commercial satellite system. The 1978 International Maritime Satellite Act designated Comsat also as United States representative to INMARSAT, a system intended to provide telex, telephone, facsimile, and data communications between ships and drilling rigs and shore.

III.5. Regional and Domestic Satellite Systems -- As originally conceived, the INTELSAT system was to provide all satellite-based international public telecommunications. From the inception of INTELSAT, however, the need for separate systems designed specifically for the needs of certain regions was contemplated, and in the past few years a number of nations have coordinated regional systems with INTELSAT. Arabsat, Marisat, and Palapa-B are examples.

A question to be considered is whether the United States' commitment to the global INTELSAT system necessarily precludes the use of domestic or regional satellite facilities for international communications. If, for example, the INTELSAT system is not capable of providing a needed service of a particular kind, or if for some purposes the use of the INTELSAT system is clearly uneconomical or impractical, would the use of alternative systems warrant support? Additional pressure to reconsider a strict policy favoring "INTELSAT satellites only" comes from the expanded activities of Comsat, applications pending before the Federal Communications Commission by a number of other carriers in the United States to use domestic satellites for international service, and a general desire to promote competition in telecommunications.

III.6. Allocation of Spectrum -- The allocation of frequencies of the electromagnetic spectrum is coordinated for international users by the International Radio Frequency Board (IFRB) of the International Telecommunications Union. The IFRB records

> the date, purpose, and technical characteristics of frequency assignments by member countries with a view to achieving international recognition . . . It examines each frequency notification received for conformity with the international radio regulations (which are drawn up and revised by World Administrative Radio Conferences) and for the possibility of harmful interference. If the examination is favorable, the frequency assignment is entered in the master register; if not, it is returned to the notifying country with suggestions regarding a solution to the difficulty. Through reference

to the master register and the Board's weekly circulars, operators worldwide have access to a listing of all registered frequency assignments. 36/

Currently, the allocation of frequency is accomplished on a "first-come, first-served" basis. That is, assignments are made to whomever first develops a need for a particular frequency and notifies the IFRB. Thus, most frequencies have been assigned to developed countries. This has caused concern among developing countries who fear that adequate space will not be available by the time they are prepared to make substantial use of the spectrum. Developing countries therefore favor the a priori approach to allocation, with assignments made in advance of demonstrated need. The controversy over methods of allocation was an issue at the 1979 World Administrative Radio Conference (WARC).

> The developing countries felt that the spectrum should be planned -- that is, divided up now, with certain frequencies allocated to members of the developing world whether or not those members could use them at present . . . The United States and some other Western nations countered the arguments for advance assignment with the promise of 'fair and equitable distribution' when the time came and assured the other nations that through ever-advancing technology and other means, they would get their fair share whenever they were ready. 37/

The issue of the method of allocation has not been decisively resolved and will continue to be a topic of international discussion in the coming decade.

III.7. Allocation of Satellite Orbital Slots -- The issue of the allocation of satellite parking slots is analogous to the issue of allocation of frequency. The

---

36/ Radio Frequency Use and Management, Impacts From the World Administrative Radio Conference of 1979, Office of Technology Assessment, Congress of the United States, Washington, D.C., January, 1982, p. 49.

37/ Oswald H. Ganley and Gladys D. Ganley, To Inform or to Control? The New Communications Network, McGraw-Hill Book Company (New York: 1982), p. 100.

number of available positions for satellites in geosynchronous orbit (satellites orbiting the earth above the equator at a velocity that renders them effectively stationary relative to any point on the ground) is limited, but potential problems of crowding are likely to be ameliorated by technological advances. Developing countries are as concerned about the future availability of satellite slots as they are about the availability of radio frequencies. Although they are not equipped to make use of it at the present time, many countries claim the "right" to a share of the geostationary orbit as a "global resource." Some equatorial states have made claims of sovereignty over portions of the geosynchronous orbit above their territories. The allocation of satellite slots will be a topic of discussion and debate throughout the decade in the United Nations Committee on the Peaceful Uses of Outer Space, and through special "Space" WARCs scheduled for the mid 1980's.

III.8. Vulnerability of Systems -- A number of governments have expressed concern that increased dependence on communications and information technologies increase their vulnerabilities. For example, centralized computer data banks and computer networks may increase the possibility of coercion by terrorist groups through threats to disrupt communications or destroy records. Other examples include the interception and misuse of data by unauthorized persons, and breakdowns for technical reasons or natural disasters.

There are methods of increasing the "robustness" of communications and information systems which would render them less susceptible to breakdown. These include building in back-up lines and additional computers for the system, (redundancy), and establishing rigorous international technical standards. Unfortunately, these remedies may entail substantial increases in the cost of the systems and hence in the cost of services to users.

The legal questions of liability for damage incurred for information lost in a system breakdown, for information appropriated by unauthorized parties, or for errors in content require attention.

III.9. Technical Standards -- The setting of technical standards for communications facilities is essentially an attempt to establish common practices that will enable components of the global communications system to fit together and operate. There are a wide range of issues -- from the physical coupling of transmission lines with switching apparatus, to keyboard log-on protocols for establishing communications among geographically dispersed computer terminals -- that have to be "standardized" in some way to promote a smoothly functioning international communications system. The variety of international standards for communications facilities is revealed by the activities of the Consultative Committee on International Telegraph and Telephone (CCITT) of the International Telecommunications Union. This organization issues recommendations on general tariff principles, radio and television program transmissions, telephone switching and signaling, telegraph switching, and new data networks to name just a few. Subgroups consider such detailed questions as standardization of symbols and other aspects of subscriber equipment to meet human factor needs, standardization of subscriber control procedures for supplementary telephone service, and standardization of tones. Two other organizations that are members of the CCITT, the International Organization for Standards (ISO) and the International Electrotechnical Commission (IEC), are "nongovernmental organizations whose sole purpose is the coordination and approval of voluntary international standards of various types, with a growing interest in telecommunications standards."[38/]

38/ The CCITT: Organization, U.S. Participation, and Studies Toward the ISDN, D.M. Cerni, Report Series of the National Telecommunications and Information Administration, 82-101, U.S. Department of Commerce, April 1982, p. 23.

IV. Services Offered; Terms and Conditions of Use of Facilities

Though not totally separable from issues concerning the structure of facilities, the issue of services offered and the terms and conditions of use of the facilities are a somewhat different question. They concern, in general, the question of access rather than of capacity and focus on the interests of users rather than service providers. Twelve issues have been selected for annotation under this heading.

IV.1. Trade in Services -- The principal "output" of the communications and information sector comes in the form of intangible services rather than physical products. These services include telephony, telegraphy, electronic mail, data processing, information retrieval from data bases, computer programming, motion pictures, news syndicates, and business and legal consulting, among many others. Most of these services may either be purchased in the home market from domestic suppliers or imported, that is, purchased in the home market from foreign suppliers operating from other countries. Domestic suppliers also find opportunities to export their services to foreign markets.

Even though international telecommunications channels are provided jointly by the domestic and foreign communication entities, use of them does constitute a trade in services. To provide a simple example, in a call from New York to Paris, part of the charge paid is the purchase of an in-country communication service from the United States carrier (from New York to the middle of the Atlantic) and part is the service imported from the French correspondent carrier (from mid-Atlantic to Paris). In theory, when this call is paid for, some of the charge is remitted to the foreign carrier by the domestic carrier and the balance of payments accounts are thereby affected. In practice, however, this swapping of charges is actually handled in bulk at accounting intervals through a settlement process between the carriers.

By far the most significant trade issues arise not from charges for telecommunications services as just described, but from the use of telecommunications facilities in other international commercial transactions. Rules and regulations on the use of such facilities thus have implications far beyond the issue of pure telecommunications revenues.

Assessing the dollar value of trade in communications and information services is no meager task. The dollar values of these exports and imports are not known with much precision, as the official balance-of-payments accounts are not set up to provide the kind of detail required to derive the values for this sector. Furthermore, trade figures based solely on communications would seriously understate the real value of these services in international commercial relationships; communications services support trade activities of all other types of services and of trade in merchandise.

In 1980, exports of services of all kinds by the United States amounted to nearly $100 billion. Nearly 40 per cent of this value was attributed to receipts of income from direct investment abroad.[39/] The significance of the operations of foreign subsidiaries of American companies (in trade and for the balance of payments) is further demonstrated by the fact that 20 per cent of total revenue of the 50 largest multinational enterprises based in the United States was derived from foreign revenue, and that foreign revenue comprised over 50 per cent of the total revenue of the 20 largest multinational enterprises.[40/] These and other enterprises rely heavily on international flows of information to conduct their business, and on reliable, cost-effective access to

39/ Survey of Current Business, United States Department of Commerce, Volume 61, Number 3, March 1981. Table 1, p. 50.

40/ "New Targets and Familiar Faces -- Special Report Spotlight and International Business," Forbes, July 7, 1980, pp. 102-103.

telecommunications facilities and information services in the countries where their subsidiaries are located. They rely both on market services provided by service bureaus, and on non-market or "intra-corporate" flows of information provided from within the corporation.

Hence, a very significant aspect of the emerging international regime that requires attention is the question of how rules and regulations regarding the terms and conditions of use of communications facilities will affect trade in services and foreign direct investment. Interests likely to be affected span the industrial spectrum. The structure and efficiency of international concerns in banking, insurance, travel, automobiles, petroleum, advertising, and many others are all affected by barriers that obstruct flows of information or by regulations on access to telecommunications facilities. The economic ramifications of such barriers are unquestionably of great importance.[41/]

IV.2. Rate Structures -- Policies and practices regarding the structure of rates and the resulting prices paid for international telecommunications services by users are significant determinants of the way information flows internationally. In the Congressional report on "International Information Flow: Forging a New Framework," a number of United States corporations describe their experiences in testimony to illustrate how excessive and discriminatory tariff and pricing practices hamper the flow of information by making operating costs prohibitive and effectively eliminating some providers of goods and

---

41/ An interesting treatment of some of the issues raised here can be found in, Joseph P. Markoski, "Telecommunications Regulations as Barriers to the Transborder Flow of Information," Cornell International Law Journal, Volume 14, Number 2, Summer 1981, pp. 287-331.IV

services from markets.[42/]

Recommendations on rate structures are promulgated with United States participation by the Consultative Committee on International Telegraph and Telephone (CCITT) of the International Telecommunications Union (ITU). The recommendations are non-binding on participants, but are very influential in establishing international practices in this area. An idea of the general direction being taken regarding recommendations on rates is revealed by selecting a few representative illustrations from General Tariff Principles; Charging and Accounting in International Telecommunications Services.[43/]

The recommendation on "Costs and Value of Services Rendered as Factors in the Fixing of Rates" begins by establishing criteria for cost-based pricing, stating that

> The income from the totality of services provided by a telecommunications organization should cover all the costs incurred by that organization, namely: (a) operating expenses; (b) interest on capital involved; (c) fiscal charges; (d) depreciation of equipment; (e) cost of research and development; (f) capital investment (as required).

42/ Thirty-Second Report by the Committee on Government Operations, December 1980, (U.S. Government Printing Office: Washington D.C.), p. 14. Examples of some of the problems noted by firms include statements that "the cost of using the public communications network is two and a half to five times more expensive in European countries than in the United States," and that "the trend of unequal increases in charges for data communications services, particularly for inter-European or intercontinental services . . . are already taking place . . . ," and "when a foreign government . . . forces U.S. users to switch from full period to usage-sensitive service, it means higher profits for the foreign entity, higher costs for the U.S. user, and higher prices to U.S. consumers." (pp. 13-14).

43/ Recommendations of the D Series, Yellow Series, Volume II, Fascicle II.1, VIIth Plenary Assembly, 10-21 November 1980, International Telecommunications Union, Consultative Committee on International Telephone and Telegraph, Geneva, 1981. Referred to hereafter as the CCITT Yellow Series.

However, the way is made clear for bringing other factors into play for determining rates; for

> political or social reasons . . . the CCITT considers that the overall balance in the telecommunication services required should be achieved by applying an increase factor to the rates of other telecommunication services in the same telecommunication organization which will compensate for the deficit incurred by services run at a loss.

This is another way of saying that cross-subsidization is an accepted practice in the determination of rates, something the United States has been trying, on the grounds of efficiency, to avoid as a pricing principle in the domestic arena.

The recommendations on charging principles for shared-use networks[44/] specify additional criteria for pricing services.

> The provision of international telecommunications facilities is subject to tariffs which take account of the volume of information transmitted, the unit of traffic measurement chosen and the origin and destination of the information transmitted. The charge may also include a fixed component, . . . 45/

For public data networks, factors which may be taken into account in developing rates include, "user class of services, optional user facilities, type of switching, volume of data and/or duration of call, distance, time (peak and off-peak periods), route, other functions;" and, "in the case of packet switching, both duration and volume of information should be the normal accounting elements."[46/]

44/ See the annotation on this subject in IV.8.

45/ CCITT, Yellow Series, Recommendation D.6., "General Principles for the Provision of International Telecommunications Facilities to Organizations Formed to Meet the Specialized Communications Needs of their Members," (Geneva, 1981), p. 17.

46/ CCITT, Yellow Series, Recommendation D.10, "General Tariff Principles for International Public Data Transmission on Public Data Networks Dedicated to This Type of Transmission," (Geneva, 1981), pp. 20-21. On the question of measuring volumes of information see the annotation V.6. on "Valuation and Taxation of Information."

These examples provide a clear illustration of the detail in which standard rules for rate structures are specified in international discussion. Issues that require attention include whether the interests of service providers and users in the United States are finding adequate and effective representation in the CCITT, especially in light of the complaints expressed in congressional hearings referred to earlier. Furthermore, are the interests of service providers and of users similar or, at least, compatible, or do they differ sufficiently to render the presentation of a uniform United States "position" on rate structures infeasible?

IV.3. Liability for Economic Loss Due to System Failure -- Among large institutional users of international telecommunications facilities there is concern that PTT's intend to gradually phase out leased-line networks. The PTT's will do this as a means of moving users onto the growing public networks. The fear is that public networks may be more susceptible to system failure, and hence economic losses would be caused by delayed, garbled or undelivered information. The issue is not merely a concern with recouping a portion of the service charge for the time the system is not functioning properly. Far more sizable losses may ensue from missed opportunities or botched transactions that rely on timely, accurate information.

According to CCITT recommendations, "administrations are not obliged to accept any responsibility for the end-to-end quality of transmissions over connected circuits which comprise a private leased circuit network."[47/] Nevertheless, a user of leased lines is prepared to assume the relatively low risks inherent in a dedicated leased-line system. They are of tested reliability and entail only small and predictable delays. Furthermore leased-line networks use

---

47/ CCITT, Yellow Series, Recommendation D.1 "General Principles for the Lease of International Private Leased Telecommunication Circuits," (Geneva, 1981) p. 8.

terminal and switching equipment specially selected for the user's particular needs. On public networks, however, where the service may not be as reliable, the user may be forced to bear a much greater risk for failures over which he has far less control. Who should be held liable should an economic loss be sustained?[48/]

IV.4. Use of Public Data Networks -- Currently, leased-line networks are available from communications administrations at flat monthly rates and can be augmented by the leasee with supplementary equipment to obtain high-quality, economical service. Public networks, however, are available at time-sensitive or volume-sensitive rates, and generally do not have the same reliability or range of services available on leased-line networks. They are less likely to meet the particular needs of individual customers.

Two issues of interest here are first, the nature of the rules that permit private lines to tie into the public networks, and secondly, the broad concern over the possibility that PTT's may seek to phase out leased lines in order to force users onto public networks. There are CCITT recommendations that address themselves to these two questions. On the use of public networks used in conjunction with international private leased lines the choice is left up to the individual administrations:

> If the national law or established practices of an Administration participating in the establishment of the services do not allow access,

---

48/ This is one example of the kinds of questions on liability that may accompany the increase in international flows of information and the evolution of the regime that controls the flows. To cite one other, the Science Advisory Board of the International Telecommunications Union has undertaken a study on issues of liability in middle and outer space stemming from collisions with abandoned spacecraft left in geostationary orbit.

the relevant Administration has the right to refuse such access on its side. 49/

The preference of the CCITT recommendations for use of public networks over private networks comes through as well.

> . . . the establishment of a private leased circuit network may be authorized to meet the specific technical and operational requirements of certain customers, if requirements cannot be met by the public networks or by specialized networks set up by Administrations . . . 50/

There is concern that as public networks improve in quality and variety of service, communication administrations will take steps to phase out leased-line services.

IV.5. Direct Broadcast by Satellite -- Television broadcasts have been transmitted by satellite since the 1960's. In the traditional arrangement the programming is first carried to a sending earth station by conventional means, then transmitted via satellite to a receiving earth station, and then delivered, again by conventional means to over-the-air broadcasting stations for dissemination to home viewers.

By comparison, direct broadcast by satellite (DBS) involves beaming the programming directly from satellite to the home where viewers receive the signals directly by means of their own small antennas. Other equipment may be necessary for converting the transmission frequencies to usable form for standard television sets.

The international issue is whether TV (or radio) programs may be beamed via satellite from one country directly to the TV sets of residents of another

---

49/ CCITT, Yellow Series, Recommendation D.1, Section on the "Use of Public Networks in Conjunction with International Private Leased Circuits," (Geneva, 1980) p. 8.

50/ CCITT, Yellow Series, Recommendation D.1 Section on "Private Leased Circuit Networks," Geneva, 1980. p. 7.

without the prior consent of the government of the receiving country. For some countries, "prior consent" would mean allowing the receiving state to examine and censor each program; for others "prior consent" would mean permitting the receiving state to license DBS service before it begins, with requests for such licenses evaluated on the basis of local standards.

Other issues requiring international agreement or discussion include selection of the frequency bands
allocated for DBS in order to prevent interferences with other services, technical standards for the systems, and a priori planning versus evolutionary development of the service.[51/]

IV.6. Value-Added Networks -- A value-added network (VAN) is one in which private leased circuits are obtained from a common carrier and augmented by the leasee with computer switching and processing devices (which create the value added) providing services of various kinds to third parties.[52/]

The CCITT draws very clear distinctions between use of the computer in a value-added network to perform "switching" services and its use to perform "data processing" services. In its recommendation on "Use of private leased circuits in conjunction with computers (data processing centers) operated by

---

51/ "Detailed planning is a list of frequency channels and orbit positions assigned to each country and includes considerations such as strictly defined service and coverage areas. The evolutionary approach simply accommodates new requirements as they arise, in accordance with the procedures that are laid out in the (ITU) Radio Regulations." Richard E. Shrum, "International Regulatory Aspects of Direct Broadcast Satellite," in Direct Broadcast Satellite Communications, Proceedings of A Symposium, National Academy of Sciences, Washington, D.C., 1980, p. 81.

52/ The Federal Communications Commission used the term "enhanced service" meaning a combination of communication and computer processing in which the computer processing is used to "act on the format, content, code, protocol or similar aspect of the subscriber's transmitted information." Second Computer Inquiry, Federal Communications Commission, Volume 77, 2d Series (1980), p. 498.

customers and providing data processing services to others,"[53/] it states that,

> Data processing involves the use of the computer for a combination of operations such as calculating data, merging, sorting, and so on, in accordance with programmed instructions, as distinguished from circuit, message or packet switching.

The CCITT recommendations permit the former type of enhancement of their facilities but not the latter, if it means transmission of messages between users.[54/]

IV.7. Interconnection of Equipment -- The variety and quality of communications and information-processing services depend, in part, on the freedom with which users are permitted to augment the basic communications networks by attaching their own equipment -- telephone, computer terminals, switch boards, encryption devices, modems, and so on.

In the United States, regulations permit the attachment of privately-owned equipment to the telephone network in most situations -- regardless of whether the equipment is produced domestically or in another country. Such liberal policies have generally not been adopted by other countries. This fact has implications for U.S. exporters of such equipment who will not find receptive markets for their hardware where interconnection is limited or confined to domestically produced products. Foreign countries' restrictive policies also affect U.S. firms operating within their borders. These companies may wish to upgrade the services of the local telephone networks with equipment available

---

53/ CCITT, Yellow Series, Recommendation D.1, "General Principles for the Lease of International Private Leased Telecommunication Circuits," (Geneva, 1981) p. 9.

54/ "Leased circuits connecting users with a data processing center may not be used for the exchange of information between user terminals either directly or on a store and forward basis." The primary concern here is expressed by the phrase "the customer shall not be permitted to operate in the manner of an Administration by providing telecommunication services to others," CCITT, Yellow Series, p. 9.

from U.S. manufacturers.

International negotiations have begun on this issue. They are being fashioned after the Agreement on Technical Barriers to Trade (Standards Code) of the Tokyo Round of the Multilateral Trade Negotiations. In December of 1980, the United States and Japan initialed a joint statement specifying principles that will ensure that foreign manufacturers are permitted access to the respective "interconnect equipment markets." The next step is to broaden the agreement to include other countries.

There is a comment on this general issue in the CCITT recommendations. It concerns shared-use networks and states that,

> In certain cases the (telecommunication) Administration may . . . require that certain equipment intended for the operation of the circuits concerned (for example, switching, concentration or multiplexing equipment) be installed on the Administration's premise and/or be provided by the Administration. 55/

And, with regard to private leased lines, that

> Equipment connected to a private leased circuit must meet the technical conditions laid down by each (telecommunications) Administration concerned. Where consultation would assist in expediting type-approval of equipment, Administrations should consult with each other. Moreover, the equipment in customer or user premises should not allow the circuits to be used in conditions other than those authorized. 56/

These examples extend beyond the narrower issue of attaching equipment to the public telephone network but involve essentially the same kind of considerations.

55/ CCITT, Yellow Series, Recommendation D.6, "General Principles for the Provision of International Telecommunications Facilities to Organizations Formed to Meet the Specialized International Communications Needs of Their Members," (Geneva, 1980), p. 17.

56/ CCITT, Yellow Series, Recommendation D.1, "General Principles for the Lease of International Private Leased Telecommunication Circuits," (Geneva, 1980), p. 4.

IV.8. Shared-Use Networks -- By the terms of prevailing international practice, users can meet their telecommunications needs either through access to public networks[57/] or by establishing their own networks by leasing dedicated lines from telecommunications authorities.[58/] For certain groups of users trying to achieve specialized commercial objectives, however, neither of these options may provide the kind of tailored and cost-effective service their activities require.

Two examples of industry consortiums that have devised special-purpose shared-use networks (essentially, networks of private leased lines shared by many different firms of the same industry) to meet the peculiar needs of their crafts are SWIFT, the Society of Worldwide Financial Telecommunications, and SITA, the Societe Internationale des Telecommunications Aeronautiques. The former provides communications services for certain financial transactions to the worldwide banking community, and the latter provides communications services for segments of the worldwide travel industry.

Portions of the CCITT recommendations refer to practices for the establishment and use of such networks.[59/] They state that

> Administrations may authorize the exceptional use of . . . facilities made available to those organizations formed to meet the specialized international communication needs of their members only if such

---

57/ See annotation IV.4.

58/ See annotation IV.10. Users may lease dedicated lines from telecommunications authorities only under certain conditions.

59/ For a discussion of some of the practical difficulties encountered by SWIFT in trying to establish its network see Thomas J. Ramsey, "Europe to the Challenge of the New Information Technologies: A Teleinformatics Strategy for the 1980's," Cornell International Law Journal, Volume 14, Number 2, Summer 1981, p. 279.

members have common interests and exercise the same activities in areas other than telecommunications. 60/

The recommendations regarding the perpetuation and sustained availability of shared-use networks, however, are discouraging in tone. They state, for example, that,

> . . . Administrations may exceptionally and subject to national legislation, make available international telecommunications facilities for a use which cannot be authorized under the provisions of other CCITT recommendations and which cannot be met by existing public services. These facilities may be provided for a period until new public services or specialized networks which meet the reasonable requirements involved become available.

This idea of eventually replacing such specially authorized facilities is again reinforced by,

> . . . Administrations reserve the right to substitute for (these) facilities made available . . . services provided by them when such services have evolved to accommodate customer requirements . . .

The issue of all communications being eventually channeled through public networks is a significant one from the users' point of view because it raises concerns that range from the quality and variety of services available, to liability for economic loss because of system failure.

IV.9. Cryptography -- Concern over unauthorized surveillance or other forms of disruption of communications have accompanied an increase in the volume and types of flow of information -- electronic funds transfer, electronic mail, and proprietary corporate data of all sorts. There is a growing demand for new or improved techniques of assuring the security of communications.

60/ CCITT, Yellow Series, Recommendation D.6, "General Principles for the Provision of International Telecommunications Facilities to Organizations Formed to Meet the Specialized International Communications Needs of Their Members, pp. 16 - 17. This citation applied as well to the remaining quotations in this annotation.

Cryptography, the use of codes to render messages or data unintelligible to unauthorized parties, is one technique.

Cryptographic methods are generally used to prevent two kinds of tampering. The first is "passive" surveillance, through which the unauthorized party gains access to the signals or messages of other parties (with or without their knowing it), obtaining knowledge of the content, which may then be used to some commercial or political advantage. The second kind of tampering is the "active" variety, when the unauthorized party not only gains access to the messages of other parties, but may then change the content, moving a decimal point a few places in a financial transaction, or changing a word or meaning in other communications.

Until recently, cryptography had been an issue of interest largely to agencies of government responsible for national security that are responsible for sending secure messages and breaking codes. A growing demand for secure communications in commercial activities, however, engendered a transfer of cryptographic technology to nongovernment users.

Two related aspects of the cryptographic issue have emerged as issues of policy. One concerns the national security implications of research on cryptography, and the broad dissemination of new cryptographic techniques. The other concerns the encryption of everyday communications. The U.S. Government's primary concern regarding the first aspect is

> that open research and publication in cryptography jeopardize national security by making available to foreign governments encryption techniques that (the National Security Agency) would have difficulty breaking, calling to the attention of foreign governments the vulnerabiity of their current encryption methods, and revealing knowledge that might endanger the inviolability of codes used by the U.S. Government. 61/

61/ Stephen H. Unger, "The Growing Threat of Government Secrecy," Technology Review, February/March, 1982, p. 32.

Controversial efforts have been made by government to inhibit the transfer of technology on cryptography.62/

The issue regarding the potential demand for and application of cryptographic techniques in international communications include the availability of cryptographic services altogether and the question of standards. Tariffed encryption services are not now available from either domestic or foreign international communications carriers. In developing such services, there will always be at least two telecommunications entities involved, and they will have to come to agreement on procedures to be followed and equipment to be used. In addition, the wide adoption of tariffed encryption techniques could require an elaborate international standards-setting procedure. Encryption services available to the users of private dedicated leased-line networks may be less cumbersome to develop, and may provide additional incentive for resisting actions of communications administrations seeking to encourage the use of public networks.

IV.10. Private Leased Lines -- Large institutional users of communications services with specialized needs growing out of their international commercial activities frequently establish private networks dedicated exclusively to their own use. They do so by leasing circuits from common carriers that provide international services and from foreign telecommunications administrations.

> The international telecommunication private leased circuit service is described as making one or more international telecommunications circuits available to a customer for his dedicated use on the terms and conditions which may be set out in a lease agreement between

---

62/ For a discussion of some of these efforts see David Kahn, "Cryptography Goes Public, Foreign Affairs, Fall 1979, pp. 141-159.

the customer and the (telecommunications) Administration of the country at each terminal of the circuit. 63/

The CCITT has promulgated recommendations on the establishment and operation of a leased-line network regarding duration of lease, charging, procedures for cancellation, collection of charges and accounting, allowances for interruption, use of public networks in conjunction with international private leased circuits and use of private leased circuits, in conjunction with computers.64/

The continued availability of leased lines is a topic of some concern. Many communications administrations are developing more efficient public networks, and may seek to have users gradually shift over to these public networks by changing the terms of availability of leased lines. In the longer term the introduction of new technologies, such as integrated services digital networks (ISDN), will provide alternatives to leased lines. The implications of a movement away from leased lines is sure to become a vital consideration in strategic planning of international firms.

IV.11. Resale of Transmission Capacity -- Users who lease private lines from communications administrations may find that after meeting their own needs they have excess transmission capacity on their dedicated networks and may want to "resell" some of that capacity to other users. They could collect revenues by charging the secondary user a fee for a portion of their unused capacity. They could do this without themselves incurring additional charges for the lines as they are leased from carriers at flat monthly rates. In international practice there are some strict prohibitions against this kind of resale, as it has

63/ CCITT, Yellow Series, Recommendation D.l "General Principles for the Lease of International Private Leased Telecommunication Circuits," (Geneva, 1981) p.3.

64/ CCITT, Yellow Series, Recommendation D.l, pp. 3-10.

the effect of channeling revenues for communications services away from the communications administrations and to the reseller. CCITT recommendations, for example, state that

> ... private leased circuits may be used only to exchange communications relating to the business of the customer. When the circuit is used to route communications from (to) one or more users other than the customer, these communications must be concerned exclusively with the activity for which the circuit is leased, (and) ... the customer may derive telecommunications channels from a private leased telephone-type circuit. These channels ... may be extended by means of other circuits leased by the same customer. The channels so derived must not be subleased ..., (and finally) Administrations shall refuse to provide an international private leased circuit when the customer's proposed activities would be regarded as an infringment of the functions of an Administration in providing telecommunication services to others. 65/

In contradistinction to these practices and prohibitions, the Federal Communications Commission in the spring of 1980 issued a notice of rulemaking that proposed the adoption of a policy that would "prohibit United States carriers from restricting the resale or shared use of international telecommunications services and facilities."[66/] There is some concern that unilateral adoption of the proposed rule by the FCC could cause a curtailment in the availability of flat-rate leased lines by telecommunications administrations opposed to resale. In any event, such a position can hardly be sustained without the agreement of the foreign correspondent carriers.

IV.12. Integrated Services Digital Network -- Integrated Services Digital Network (ISDN) is a concept touted as the communications network of the

---

65/ CCITT, Yellow Series, Recommendation D.1, "General Principles for the Lease of International Private Leased Telecommunication Circuits, (Geneva, 1980) p. 4.

66/ Joseph P. Markoski, "Telecommunications Regulations as Barriers to the Transborder Flow of Information," Cornell International Law Journal, Volume 14, Numbers, Summer 1981, p. 322.

future. It is conceived of as a "general purpose" network in which all forms of communications -- voice, data, record, and video -- are converted to digital code, essentially a stream of bits, that can be conveniently and efficiently manipulated by computers serving as "intelligent" switching devices. Thus all types of communications can be conveyed over a common digital network. The development of international standards for ISDN, and the establishment of the terms and conditions of use of these networks, will likely be a significant aspect of the evolving communications "regime" in coming years.

V. Information Per Se

The following eight issues pertain to the treatment of information itself, as opposed to the facilities or particular kinds of services. The issues range from those that have been part of international debate for many years, such as press freedom and the treatment of journalists, to issues that have emerged as a direct result of recent technological advances, such as remote sensing by satellite. Among the topics covered, the treatment of information is probably the least developed in terms of understanding and established practice.

V.1. Press Freedom -- In the area of the mass media, particularly print and broadcast journalism, there have been growing efforts by individual governments and international organizations to impose restrictions on the activities of journalists. This issue has arisen as part of the debate on the "new world information order." This is a rallying cry used by some representatives of developing countries for a reduction of Western dominance over the mass media and a redressing of what they view as an "imbalance" of flows of information.

At its Belgrade Conference in November of 1980, the UNESCO Secretariat was authorized by a majority of its membership to draft a document on the "new world information order." A draft of the section regarding the press included provisions that would require journalists to subscribe to a code of ethics on information and to write favorably about governments' policies in countries in which they are stationed. Journalists would be issued identification cards which could be withdrawn by governments in retribution for unsanctioned behavior.

In reaction to these proposals, a "Voices of Freedom" Conference, organized by private groups involved with the mass media, was held in May of 1981 in Talloires, France, to express unqualified support for press freedom and to oppose the UNESCO proposals. The United States Congress introduced resolutions stating that the position of the United States regarding freedom of information is non-negotiable and threatened to suspend funding or to withdraw its membership in UNESCO altogether if the organization pursued its attempts to control the flow of information. Although the issue has become less volitile, it has not disappeared.

A separate consideration on this subject of censorship, control of information and press freedom involves the possibility of governments manipulating the networks that transfer information. If governments attempt to control flows of "economic" information through restrictions on the location of data processing or by surveillance schemes to permit taxation, severe impediments may also be imposed -- intentionally or unintentionally -- on the information of the mass media. The disruption of mass media information may well be a side effect of government schemes to impede, value, or measure information flows for "legitimate" reasons of economic policy.

V.2. Privacy Protection - Natural Persons -- With the advent of computers and their capacities for storing and transferring large amounts of information, protecting the privacy of individuals has become a paramount concern. Practices followed by financial and insurance institutions, retailers, employers, and governments regarding the kinds of information collected on individuals and the way it is stored and used has been the subject of attention by many governments.

To protect personal data stored in automated systems, several European countries including Austria, Denmark, the Federal Republic of Germany, Norway, Sweden, and Luxembourg have enacted comprehensive or "omnibus" privacy laws uniformly covering all sectors of the economy. There is a comparable law being proposed in the United Kingdom.

The United States, however, has not followed this European approach of having a single omnibus law covering all sectors of the economy, and has not created a "data commission" to oversee the registration of data bases. Instead, privacy law in the United States is derived from a variety of sources, including the Constitution, common law, and statutes and regulations at both Federal and State levels.[67/]

While governments can work to improve privacy protection for their citizens through their domestic laws, they have far less control over the treatment of information held in computers outside of their own territories. The communications and information technologies that permit large volumes of information to flow rapidly from computers in one country to computers in another, compromise the effectiveness of data protection laws unless the receiving countries have comparable safeguards. Judging what constitutes "comparable" safeguards among countries with different legal traditions is not a simple matter.

As a means of avoiding the erection of barriers to flows of personal data

---

67/ A thorough treatment of the approach to privacy protection taken in the United States is provided by Robert Aldrich in "Privacy Protection Law in the United States," Report series of the National Telecommunications and Information Administration, U. S. Department of Commerce, 82-98, May 1982.

among countries in the name of privacy protection, there have been international efforts to establish commonly accepted standards for the protection of individual privacy. The "Guidelines Governing the Protection of Privacy and Transborder Flow of Personal Data," a voluntary agreement signed by the United States and adopted by the Organization for Economic Cooperation and Development (OECD) in September of 1980,[68/] is the result of one of these efforts.

The Guidelines outline minimum standards of privacy protection, including:

- limitations on the collection of personal data,
- requirements that what is obtained be relevant, accurate, complete, and up-to-date,
- specification of the purposes for which the data will be used,
- limitations on disclosure without the subject's consent or by authority of law,
- safeguards against unauthorized access,
- openness about developments, practices, and policies with respect to personal data, and
- the right of subjects to see information about them and to challenge, correct, or amend it.

The OECD Guidelines have provided a basis for voluntary agreement among countries with various national approaches to the protection of privacy.

In October of 1980, the Council of Europe with 21 member nations[69/] adopted a "Convention for the Protection of Individuals with Regard to Automatic Processing of Personal Data." When ratified by five member nations, it will be a treaty, legally binding on the countries that have ratified it. The United States, however, would not ratify the Convention without unprecedented legislative changes. These changes would include the enactment of a statute covering all computerized personal data held privately or by the government, as

---

68/ Twenty-one out of 24 member countries of the OECD have endorsed the Guidelines as of this writing.

69/ The United States is not a member.

well as the requirement that data bases be registered with a central authority. The United States does not approach the issue in this way.

There has been some fear that privacy protection would serve as the ostensible rationale for inhibiting international flows of information with the real motivation being commercial protectionism. To date, this fear seems to have gone unrealized.

<u>V.3. Privacy Protection - Legal Persons</u> -- In most instances, laws and practices regarding privacy protection pertain exclusively to individuals, or "natural persons." However, the data protection laws of Austria, Denmark, Luxembourg, and Norway extend the same protections to corporations and institutions, or "legal persons." Such laws increase the burden of compliance for international firms and may inhibit flows of many kinds of data necessary to conduct business internationally.

<u>V.4. States As Data Subjects</u> -- A third category of "data subject" in addition to "natural persons" and "legal persons" is the state or government. Some governments may declare "sovereignty" over information about them and their nations, thereby necessitating the granting of "prior consent" before information about that country can be transferred abroad. This issue is raised, for example, in the area of remote sensing by satellite[70/] when some governments claim sovereignty over information about their territories picked up by satellites. This concept of "prior consent" applied to current events could have a serious effect on journalistic freedom.

<u>V.5. Property Rights on Information</u> -- Incentives to create new information such as contents of data bases, innovative computer software programs, novel "firmware" programs with algorithms permanently etched on

---

70/ See annotation V.8.

microchips, are provided in part, by the expectation of financial returns from selling the information. Some of the peculiar attributes of information, however, complicate the enforcement of property rights for this intangible "commodity" through patent and copyright laws.

Factors which make the enforcement of property rights a problem include the following: Information does not come in well-defined units, it is not "used up" in consumption, the creation of information can be costly, but the incremental cost of reproducing it very low, and excluding nonpaying beneficiaries is very difficult. While computer and communications technologies have created new opportunities for creating and disseminating information, they have simultaneously complicated the methods of establishing and enforcing property rights. Important questions to consider are: How will these technological developments and the accompanying difficulties in enforcing property rights affect incentives to produce new information? What additional international agreements or procedures should be promulgated to strengthen and protect property rights on information? Would such international agreements accommodate attempts by governments to place a value on information for the purpose of taxing it?

V.6. Valuation and Taxation of Information -- Under certain limited circumstances, information can be packaged and sold just as any other "commodity," with its market value established by transactions between sellers and purchasers. The market value is determined essentially by prevailing competitive and technological conditions that affect the supply of information and by what the purchasers believe the information will be worth in their use of it. Their assessment of its value is expressed by what they are willing to forego to have the information, which is indicated by their spending a particular amount of money on the information instead of on something else.

If they have the authority to do so, tax agencies may conveniently assess some sort of excise tax on the basis of the transaction, just as might be done with any other commodity.

A vast majority of information, however, while certainly of great value to its user, never has an "objective" market value established for it. Information that serves some "valuable" intermediate function in the production process of a firm by contributing to the ultimate value of its final product has no "market" value as described above, and hence no objective basis upon which to assess a tax.

The international aspect of the value of information comes into play when information, which is an intermediate component of the production process, is transferred among the geographically dispersed subsidiaries of a multinational firm. Officials of a particular country may consider that something of value is crossing the border, and that it ought to be subject to the same tariff procedures or customs duties as other objects of value crossing the border.

There seems to be a growing concern that an increasing proportion of the value of economic activity in the future will emanate from the production, processing, and dissemination of information, and that the provision of information-intensive "services" will gradually overtake production of physical merchandise as proportions of total output in developed economies.

If the economies of the future are information-based, governments will have to ascertain how they can maintain an adequate tax base. In the case of countries that rely on the value-added tax, the methods of assessing taxes will require attention. When an increasing proportion of value added comes from information with no objective market value, how can the taxing authorities efficiently distribute the tax burden?

A second motivation for attempting to place a value on information, which then becomes taxable, arises from a perceived increase in the volume and significance of international information flows, primarily in the form of computer-to-computer communications among dispersed subsidiaries of transnational corporations. With information entering and leaving countries in this manner, a great deal of "value" is crossing the border. Some governments would argue that the information ought to be subject to the same tariff principles that apply to physical goods that have value and cross borders.

Placing a value on information for purposes of taxing or tariffing can have serious ramifications, ranging from the unwarranted surveillance of content (with all of the issues of privacy and political or economic abuse such surveillance would raise), to the introduction of serious inefficiencies in the production processes of individual firms. Information is largely an intermediate product used by firms to produce other goods and services. There is no reliable way to assign value to information at intermediate points in the production process. Even knowledge of the value of the final products that "embody" the information will not permit non-arbitrary assignment of value to the information itself.

Two separate concepts pertaining to measurement often confused with one another are: (1) measuring the value of the <u>use</u> of information; and (2) measuring the arbitrary, symbolic representations of information, such as pictures or words or bits or bytes. The second concept also concerns the rate and duration of the flow of such symbols (or their electronic counterparts) over communications networks and facilities. It is the former concept of establishing the value of the <u>use</u> of information that would have to be applied in order to levy taxes on information without introducing serious and unpredictable distortions into the production process. Yet value of the use of information cannot be reliably known until the service or product embodying the information finds a buyer, and hence, a price in the market, long after it has crossed a border or has been "used" in the production process. The latter concept of mechanical tabulation, though frequently conceived of as providing measures of the volume of information flow, has no consistent or logical relationship at all with the content or with the value of information.

Recommendations on assessing the volume of information by measuring symbolic representations have been devised by the CCITT which has considered "the need to define a unit measuring the volume of information transmitted in an international packet-switched data transmission service." It unanimously declared the following view:

> (1) definition of segment: the measurement unit used for charging for the volume of information transmitted in a packet-switched service is independent of the maximum packet length. It is called a segment and its length is 64 octets (This number is subject to further study);
>
> (2) operating rule: the quantity of chargeable data contained in each message is expressed as a number of segments and is rounded off to the next higher unit. 71/

V.7. Surveillance of Content -- The issue of surveillance of content will probably become prominent if attempts are made to sharpen the techniques of valuing information flows for purposes of taxation, and as regulatory agencies implement data protection laws or impose selective restrictions on information flows for economic or social

reasons. Regulating entities must have some means of enforcing transborder data flow regulations and policies. In an article entitled "Transborder Data Flow

---

71/ CCITT, Yellow Series, Recommendation D.12. "Measurement Unit for Charging by Volume in the International Packet-Switched Data Transmission Service," p. 24. As I argued above, there is no logical or consistent relationship between "64 octets" and the value in use of the information represented by that "segment."

Regulation: Technical Issues of Legal Concern," Eric Novotny states that, "in addition to conducting audits, holding security inspections, and enforcing requirements of regulated entities, authorities can monitor the data traffic itself."72/ He notes that "technical proposals have been advanced . . . to identify data flows with some type of digital 'license tag' to identify the source, content, value, and destination of such flows."73/

Yet because of the growing variety of communication channels available, the difficulty of deciphering streams of "bits" of digital transmissions, and the availability of cryptographic techniques, some argue that technology will render attempts at monitoring content impractical. Even if this were the case, however, laws or regulations on content may have to be dealt with. Circumvention by means of technological gadgetry may bring stiff penalty.

In addition, development of the technology of circumvention may merely encourage further development of the technology of surveillance. The technology argument can be used to support both cases. Attempts at monitoring content could be one of the most serious threats to free international flows of information.

V.8. Remote Sensing by Satellite -- By use of sophisticated photography techniques and sensing devices and processing of images received, specialized satellites can gather a great deal of information -- meterological, agricultural, environmental, and geological -- for any region of the world .

Information gathered by remote sensing does not yet have a large number of private users. At the present time, information obtained in this manner has been of more interest to the public sector. The United States Government has

---

72/ Computer/Law Journal, Vol. III, No. 2, Winter 1982, p. 120.

73/ Novotny, p. 109.

developed a policy of "sensing" the territories of other countries without prior approval, and openly disseminating the data thus collected without the consent of the involved country. This is a policy which has been contested by many other countries. The Soviet Union is concerned about the possible military applications of the information, and some developing countries are perturbed about the economic power it confers on those who have the technology and knowledge to interpret and use the information. Such countries consider the attaining of such information and its dissemination to third parties an infringement of their sovereignty.

The scope of the definition of remote sensing has not yet been determined internationally. The United Nations Legal Subcommittee of the Committee on the Peaceful Uses of Outer Space is in the process of developing a set of principles on remote sensing which may have important repercussions on the evolution of this technology.

In the meantime, remote sensing has been used to further economic development through predictions of agricultural yield, and determination of mineral locations and other resources.

VI. Effects of Use or Application of Communications and Information Technologies

Column VI of Table 1 lists thirteen "effects" of the use or application of communications and information technologies. These are the issues which are affected by the nature of the evolving international regime of information flow, serving as motivations or rationales for the postures and positions taken by various governments on all other issues in the table. The wide scope of these issues covers political, economic, technological, and societal concerns.

VI.a. Freedom of Expression -- While the "right" to free expression is a concept that most democratic societies support and will work to preserve, it is by no means a universally accepted tenet. One of the primary concerns with the development of rules and regulations for components of the evolving communications regime is the effects of these rules on the freedom with which people in countries around the world can express their ideas, opinions and views, and the freedom with which others can have access to those expressions. Policies of governments devised in response to concerns about the economic and strategic significance of information and of new technologies may have unintended suppressive effects on freedom of expression. The advancing technologies bring with them both the capacity to support free expression through an increase in communications channels and diversity of content, and the potential to control or censor the content by increasing the means of monitoring or manipulating the channels. A few related observations are given on this subject in the next annotation.

VI.b. Free Flow of Information -- The concept of free flow of information has served as the principle upon which United States policy in international communications is based.[74/] Judging from discussion on the subject, however, it

---

74/ A report on international communications policy submitted by the Department of State to Congress states that "Since the end of World War II, the United States and its democratic allies have strongly supported the free flow of information, as expressed by Article 19 of the Universal Declaration of Human Rights, which asserts the right of everyone to 'seek, receive and impart information and ideas through any media and regardless of frontiers.'" Report Submitted to Congress Pursuant to the Foreign Relations Authorization Act, Fiscal Year 1979, 96th Congress, 1st Session (U. S. Government Printing Office, Washington: 1979) p. 79.

An informal document distributed by the State Department lists seven basic objectives of international communications policy, one being to "enlarge acceptance of the principle of free international flow of information and ideas, including applicability of the principle to newly emerging communications and information technologies." William C. Salmon, "International Aspects of Communications and Information," U.S. Department of State, February 12, 1982.

is clearly a concept with many interpretations. There are a number of meanings of "free flow of information:"

(1) As an extension of the First Amendment prohibition against laws "abridging the freedom of speech, or of the press," and in the sense expressed by Article 19 of the Universal Declaration of Human Rights; 75/

(2) As a condition characterized by the absence of impediments imposed inadvertently as a consequence of regulatory or technological conditions placed on the processors of computerized information or on communications facilities that transmit information.

(3) As a condition characterized by the absence of laws or regulations that intentionally impose restrictive conditions on the location and use of data-processing facilities, or on the transmission beyond borders of certain kinds of information. The motivations may be purely economic and not political, and content may be of no particular concern to the regulators.

(4) As a condition characterized by the absence of governmental attempts to require disclosure -- for economic, social or political reasons -- of the content of information being processed or transmitted.

(5) As a condition characterized by availability of information without direct cost to the recipient. 76/

Most democratic societies agree that the first definition of the free flow of information should apply to the print and broadcast media. There is less consensus about a suitable definition for entertainment and television programming, in which concerns with cultural sovereignty begin to weigh against concerns with free flow. There is even less consensus about the free flow of

---

75/ See footnote 74 above.

76/ This variety of meaning of free flows can be attributed in part to the peculiar characteristics of information. Under certain conditions it assumes the attributes of an intangible "commodity" with market-determined economic value -- a product for personal consumption or a resource in the production process. Under other conditions it conveys fundamental beliefs or rudimentary ideas to which economic value cannot be objectively assigned, and for which regulations brought to bear to achieve economic goals may clash head-on with basic rights to free thought and expression.

information in point-to-point communications, in which information is viewed as embodying more economic value than social value. If the value of information is primarily economic, it may then be measured and valued, taxed or otherwise regulated as any other "commodity," thus imposing impediments to its flow.

Some countries have reserved unqualified support for the principle of free flow as they claim not to be certain of the meaning of that concept in these various contexts.

Among the thirteen topics listed as effects of the use of communications and information technologies, this issue of free flow of information in each of its various meanings is most directly affected by the evolution of the regime. In fact, the free flow of information is the primary subject of this evolution.

VI.c. National Security -- Communications and information technologies play a vital role in the maintenance of national security and defense. Direct military applications of computers and information-processing techniques, communications networks secure from breakdown or intentional disruption, satellite systems for surveillance, cryptographic devices and techniques, and early warning devices are all essential elements of a nation's security. In formulating policies and positions on the issues of Table 1, each nation is motivated by considerations of national security. Positions taken on any international communications issue are usually qualified by the phrase "subject to the requirements of national security . . . "

VI.d. Balance of Flows and Developing Countries -- There is simmering disquiet among some developing countries concerning the global distribution of communications facilities, information-processing technologies, and information itself. In the perceptions of developing countries, the technological and economic advantages of the developed countries give the latter overwhelming command of the means of producing and disseminating all forms of information

-- current events and news, scientific and technological information, business and commercial information. According to this view, the imbalance results in one-sided news reporting by the Western press and a one-way flow of information in general, from developed to developing countries, most of which reflects the mores and cultural values of the originating countries.

Efforts to ameliorate this imbalance have been undertaken in the name of a "new world information order," a slogan which urges a redistribution of information resources -- including technological knowledge, scientific and technical information, frequency spectrum, and control over news flows -- from the haves to the have-nots.

VI.e. Societal Adjustments to Information Technology - Home and Workplace -- The broad question of the way the population in general will adjust to the use of computers and communications in everyday activities is an issue that is probably more relevant for discussions of domestic policy than for international policy, but which is affected to some extent by international developments. There are systems that bring new forms of information into the home via the television, some permitting two-way interactive communications. Home computers have been made affordable by advances in microcircuitry, and are used for all sorts of recreational purposes.

In the office, word processors are changing the nature of clerical and professional work. Robots are doing the same in the factories. Electronic funds transfer permits financial transactions to be conducted through the nearest telephone, and microchips are assuming a growing role in all sorts of gadetry, from automobiles to washing machines. Societal adjustments to these developments and their implications for health, happiness, productivity and the general quality of life are subjects of deliberate policy-making in some countries.

VI.f. Vulnerability of Societies -- The issue of vulnerability of societies is a concept similar to but of broader scope than the issue of vulnerability of systems already covered in annotation III.8. above. Societal vulnerability refers to the risks associated with becoming increasingly dependent on computers.

In 1977 the Swedish Ministry of Defense established a commission to investigate the vulnerability of computerized society and to propose measures to reduce it. Vulnerability factors in Sweden were divided into two main categories: Factors external to the computer itself -- natural catastrophies, accidents, criminal acts, misuse for political purposes, terrorist attacks, and acts of war; and factors inherent in the actual use of computers -- the concentration of computer operations, data bases that store information of a personal or confidential nature, functionally sensitive systems, and dependence on assistance from abroad.

The findings of the Swedish Commission regarding dependence on foreign sources emphasize the issues that will affect international policy development. The Commission found that

> the need for components, spare parts, and services from abroad makes . . . (Sweden) dependent on an undisturbed flow of trade . . . There are reasons to suppose that the present dependence will increase . . . (and) . . . The growing international data flow involves security and vulnerability problems of other dimensions than those existing under purely national conditions. If data processing is done on a computer in another country or continent, and if input and output data are to pass through several countries various kinds of risks increase. Protection against events abroad is naturally more difficult than to build up a domestic system of protection. 77/

77/ Ulrika Tengelin, "The Vulnerability of Computerized Society," Proceedings of the High Level Conference on Information, Computer, and Communications Policies for the 80's, Organization for Economic Cooperation and Development, Paris, October 6-8, 1980, p. 359. The general conclusion of the Commission was that the vulnerability of Swedish society is unacceptably high, and that future developments will lead to increasing vulnerability unless countermeasures are taken.

These kinds of concerns, which are "effects" of the applications of computer technologies, will probably exert an important influence on future developments in international communications.

VI.g. National Sovereignty -- The issue of national sovereignty is essentially a political matter involving two components. First, a sovereign nation has certain territorial and extra-territorial claims which it is willing to defend against encroachment by other nations; and secondly, many nations assert the power of their sovereignty in deciding what will be seen, heard, or consumed within their own borders, free from outside influence.

A number of recent developments in international communications which have intensified the concern of some governments regarding perceived threats to sovereignty include remote sensing by satellite, computer-to-computer communications across borders, and direct broadcast by satellite. Countries that lay sovereign claim to information about themselves may insist on prior consent before a journalist submits a story for international distribution, or before another country gathers remotely sensed data. Because computers and communications facilities together afford the capacity to store, process, and transfer huge volumes of information among countries without the control, knowledge, or "consent" of governments, some nations would claim that their sovereignty is threatened.

The issue of direct broadcast by satellite is a point of contention in regard to the question of sovereignty over what will be seen, heard, and consumed within the country.[78/] When broadcast "footprints" intended for one country unintentionally spill over into another, some governments will consider it a violation of their national sovereignty and will seek to establish the principle of

---

78/ See annotation IV.5.

prior consent.

VI.h. Cultural Diversity -- Free and open international communications over a variety of media increase a country's access to foreign cultures, thereby contributing to a process of greater understanding, enrichment, and diversity. The same open communications regime, however, threatens to erode traditional cultural values and beliefs and to replace them with foreign values and ideas. "Cultural imperialism" is the phrase sometimes used to describe the process by which one society imposes its customs and beliefs on another. The implication of such sentiments for press freedom, free flows of information, and open trading markets are unmistakable.

VI.i. Development of Indigenous Communications Industry -- Ensuring the healthy development of an indigenous communications industry is the basic motivation underlying many governments' positions on communications issues. To achieve greater national security, economic growth, international competitiveness and general well-being, most governments pursue policies that will assure the development of their own national capacities.

VI.j. Growth and Development -- The effects of the application of communications and computer technologies on economic growth and development is an issue of concern to developed and developing countries alike. For both, the issue revolves around methods of applying these technologies to commercial, industrial, and agricultural activities to increase productivity, output, and competitiveness. The emphasis for developed countries is discovering how communications and computer technologies can best contribute to the growth of existing industries, as well as to the development of new processes and industries. The emphasis for less developed countries for which much of the technology must be imported, is discovering the best means of access to the technologies and adapting them to their own particular needs.

Countries' needs vary with the level of development.

A concern that developing countries have is that the new information technologies may tend to widen the gap between the developed and developing world, which explains many of the views and positions taken by developing countries on components of the evolving international communications regime.

VI.k. Productivity and Employment -- Technological changes in communications, information processing, and micro-electronics[79/] have provided the means of improving productivity in virtually all sectors of the economy; but at the same time these technological changes have rekindled concerns about widespread technological unemployment.

The chain of events that would lead to technological unemployment are summarized in the following statement:

> The hypothesis underlying forecasts of vastly increased levels of unemployment is fairly simple. Micro-electronics will be a great deal more pervasive than almost any historical example of technological change, particularly since it will affect processes and products in both the manufacturing and service sectors. It results in reduced labor (and other) input requirements per unit of output, and since these productivity improvements will be very large in many instances, they are unlikely to be fully compensated for by increased outputs. Aggregate employment will fall, and this coupled with a continuing increase in the size of the labor force will result in widescale unemployment (or enforced leisure). 80/

The compensatory effects of technological change that work to increase employment levels include increased investment in capital goods, higher profits, increased real earnings, and the development of new products and services.

79/ Micro-electronics refers generally to products and activities involving the use of silicon "chips" the size of a finger nail with circuitry that contains sophisticated computer programs and stores large amounts of data.

80/ B. Boatwright, "The Employment Consequences of Rapid Technological Change," Proceedings of the High Level Conference on Information Computer and Communications Policies for the 80s, Organization for Economic Cooperation and Development, Paris October 6-8, 1980, p. 222.

Although there is concern about the effects of these technologies on employment levels, the majority of the evidence available from studies of the subject seems to suggest that predictions of abrupt and widescale technological unemployment are unsupported. There are, however, problems of technological displacement that affect particular workers or professions.

The advent of computer-to-computer communications among countries has added an international flavor to this question of employment. There is some concern, for example, that information transmitted from an in-country subsidiary of a multinational corporation to computers located in its home country for processing means a loss of jobs for the domestic economy. There has not been very much systematic analysis of this question, but it has been raised as a concern by the Canadian government.

Employment effects remain an important consideration in understanding the motivations for some of the policies adopted by countries on international communications.

VI.1. Balance of Trade; Terms of Trade -- Tariff and non-tariff barriers to trade in communications and information are restrictive methods created by governments attempting to tip the balance of trade and the terms of trade in high-technology products in their favor.[81/] While the balance of trade is defined as the difference between the value of a country's exports and the value of its imports, the terms of trade are the amount of domestic goods or services that have to be given up or paid to obtain a unit of an imported good or service. Although frequently unacknowledged, concerns with both of these economic

---

81/ Issues surrounding both trade in merchandise in computers and communications equipment and trade in services have already been covered (annotations II.2 and IV.1., respectively) and need little elaboration here.

factors underlie the rationale for government policy on issues in international communications. As international communications policies evolve, trade restrictions remain a constant hazard.

Information technologies may shift the structure of international commercial relations away from traditional export and import patterns and toward the establishment of foreign subsidiaries through foreign direct investment.[82/] Unlike physical merchandise which can easily be produced in one country, shipped to another, and then consumed at any time, intangible services often have to be provided near the purchaser, and are there immediately "consumed." Thus, domestic suppliers of services need to locate subsidiaries in foreign markets in order to provide services near where they are consumed, rather than exporting the services in the traditional sense.

For both services and goods, communications and information technologies accommodate the establishment of foreign subsidiaries by permitting fast, reliable communications among the geographically dispersed components of a firm.

VI.m. Foreign Direct Investment -- Foreign direct investment pertains to the operations of multinational enterprises.[83/] When trade barriers are in place, where cost savings can be gained, and when a high degree of management control is sought, direct investment is often preferred to exporting or licensing of patents and franchises.

82/ See the next annotation.

83/ "U.S. direct investment abroad exists when one U.S. person (U.S. parent corporation) has a direct or indirect ownership interest of 10 percent or more in a foreign business enterprise (foreign affiliate). U.S. MNC's ¼multinational corporations¢ are U.S. companies that have direct investment abroad; an MNC consists of the U.S. parent and all of its foreign affiliates." Obie G. Whichard, "Employment and Employee Compensation of U.S. Multinational Companies in 1977," Survey of Current Business, February 1982, p. 37.

In the decision-making process of multinational firms, the availability of communications services and of information-processing technologies is a factor of growing significance, especially with regard to plans for structure, location, organization of activities, and choice of products and services to be produced and marketed. The international climate in which direct investment takes place is determined by the policies and regulations of host governments regarding availability, cost, quality of service, terms of access to communications facilities and equipment, and the conditions placed on the use of information itself. This international climate and the degree of uncertainty and risk that prevail will play an important role in determining the total volume of foreign direct investment, the size of firms that operate internationally, the types of goods and services that will be provided, and the location and distribution among geographically dispersed subsidiaries. It has been suggested that businesses respond more quickly to disincentives than to incentives, since the former may signal long-term obstacles to successful investment. Both political instability and regulatory unpredictability are strong disincentives.[84/] Uncertainty about the imposition of rules or regulations on communications facilities or flows of information is another example of "regulatory unpredictability" faced when deciding to invest in a foreign country.

Advantages conferred upon multinational firms by communications and computer technologies and the implications of these advantages in terms of the control and bargaining position enjoyed by the multinational companies have become a source of concern to some governments, which if magnified, could be perceived as a threat to national sovereignty.

84/ "Relationships of Incentives and Disincentives to International Investment Decisions," Response of the Business and Industry Advisory Committee to the OECD Committee on International Investment and Multinational Enterprise, January 1981 (Updated April 1981), p. iii.

Legal Issues

Communications and information processing can raise a number of legal issues which include liability, property rights, regulation, contracts for computer services, etc. Although each country devises laws on these issues to govern its domestic activities, approaches vary among countries primarily because of differences in historical development and traditions. Legal issues can then become international issues whenever countries engage in communications or commerce, and problems arise for which methods of resolution are inconsistent. Such legal inconsistencies raise the "pure" legal issues of choice of law and territoriality.

Contracts negotiated between parties to an international transaction can clarify in advance which country's laws will apply to the transaction and which country's judicial system will have jurisdiction over the settlement of any disputes. The choice of law may be determined by assessing which country is most affected by and has the greatest interest in regulating a transaction. Nevertheless, disputes may arise over the validity of contracts that have been drawn, raising the issue of territoriality -- which country's laws should apply, or whether a country's domestic law can be applied to problems arising in foreign countries.

The experience of the Control Data Corporation (CDC) with Japan's international communications carrier, Kokusai Denshin Denwa Co., Ltd. (KDD) provides an interesting example of a territoriality dispute. KDD provided an international leased circuit to CDC on condition that it be connected to a single computer system in one location in the United States. This restriction was devised by the Japanese to prevent message switching in violation of Japanese domestic law. The arrangements made by CDC for distributed processing in the United States were permitted under U.S. domestic law. The Japanese, however, asserted in effect that CDC's arrangement undermined Japanese objectives in domestic communications regulation, despite the fact that CDC's arrangement was arguably more closely associated with interests of the United States.

There are a few general observations to make here regarding territoriality.[85]

(1) Most territorial issues are not resolved by international bodies, but by the courts and other adjudicative bodies of specific countries. Courts are bound to apply the rules of the legal system in which they are established. Therefore, the rules and principles by which the territoriality issue is decided are not primarily rules of international law, but are part of the law of the country in which the case is brought.

(2) If a statute of the country in which the case is brought is clearly intended to cover the case, that statute will almost always be applied.

(3) If the case involves a dispute between parties to a contract, and the applicable law is specified in the contract, then that choice will usually be honored.

(4) In other circumstances, the court may undertake some form of "interest analysis." It tries to balance the interests that various countries might legitimately claim in the subject matter of the case. However, a general bias exists in favor of applying the law of the forum, unless there is a good reason not to.

(5) Certain kinds of statutes -- criminal and tax laws, and complex regulatory schemes such as antitrust and securities laws -- will not be applied by foreign courts. In addition, administrative bodies rarely, if ever, have authority to apply anything but their own country's regulations. In such cases, the adjudicator does not choose between two competing laws; the ruling body either applies its own law or dismisses the case.

In summary, the territoriality question in any given case is settled under the law of the country in which the case is being decided. The decision may have been made beforehand, by statute, in which case the adjudicator has no choice

---

85/ These observations and much of the background for this discussion of legal issues were provided by Robert Aldrich of the National Telecommunications and Information Administration, Washington, D.C.

but to apply that statute. Or it could have been left ambiguous, in which case the adjudicator must interpret what he thinks are his country's choice-of-law principles. There may even be a formal or informal international agreement on the subject; but here, too, it is up to the adjudicator to decide how much legal effect that agreement has or should have in his country on the resolution of the territoriality issue in that case.

When formal or informal consensus cannot be achieved, it is usually for one of three reasons. The transactions may be so convoluted and involve so many countries that it is difficult to separate the interests involved; the law may be underdeveloped or in a state of change owing to the novelty of the activities to which it applies; or the policies underlying the law in each country may be so different and so strongly held that no country is willing to give ground to the other.

Superstructure of the Regime

Through an international process of negotiation, countries can attempt to reduce the potential for conflict in transactions undertaken for mutual benefit by harmonizing domestic laws or through extra-national agreements and treaties. "Harmonization" refers to formal or informal agreements on making domestic laws among countries more alike.

This process can thus be viewed as the legal equivalent of attempting to establish technical standards for equipment and facilities. Agreements and treaties, bilateral or multilateral, broad and sweeping in scope, or narrow and specific, have served international comity in other areas of international relations and commerce, and may play an increasingly significant role in international communications.

The questions of contract, choice of law, and territoriality underlie the day-to-day activities and developments in international transactions involving communications and information. They serve to mediate the factors listed in the lower three bars of Table 1 -- (B) decisions and practices of private entities, corporations, producers, users, indigenous and foreign; (C) actions of domestic law makers and regulators; and, to some extent, (D) intergovernmental understandings, agreements, treaties. These factors -- agreement on contracts, choice of law, and territoriality -- applied to each of the issues listed in the columns of the table constitute the "superstructure" of the evolving international regime of information flow.

APPENDIX B

Long Range Goals in INTERNATIONAL TELECOMMUNICATIONS AND INFORMATION* An Outline for United States Policy

Selected Excerpts from a Report of The National Telecommunications and Information Administration to Congress (NTIA)

NTIA Special Study Staff

Kenneth W. Leeson, Director
Robert F. Aldrich
Melvin Barmat
Bohdan Bulwaka
Jack E. Cole
Richard J. Feldman
Richard M. Firestone
Richard J. O'Rourke, Jr.
Richard D. Parlow
Kenneth G. Robinson
Roger K. Salaman
Suzanne R. Settle
Helen A. Shaw
Richard H. Shay
Francis S. Urbany
William F. Utlaut
Charles K. Wilk
Marilyn I. Wilson

Consultant Staff

Paul L. Laskin
Harrison H. Schmitt

Support Staff

Lois H. Adams
Denise F. Becker
Lorraine A. McAllan
Steven W. Pomerantz
Phyllis M. Littleford

*This appendix report to "International Communications: Blueprint for Policy" by Kenneth Leeson reflects the views of the Assistant Secretary of Commerce for Communications and Information (NTIA's Administrator) and not necessarily those of Kenneth Leeson.

Prologue

DISTURBING TRENDS, U.S. INTERESTS, AND THE NEED FOR ACTION

United States economic, defense, and political interests in international telecommunications and information services have become increasingly vulnerable to adverse foreign actions as a consequence of events over the past decade. Steps have been taken by both developed and developing nations to restrict the free flow of information across their borders. Japan, France, West Germany, Canada and other countries today are successfully targeting specific sectors of our telecommunications and information industries, generating intense subsidized competition here while imposing protectionist restrictions against American producers. The economic strengths conferred by sheltered foreign markets are in fact being used to deny American firms a full and fair chance to compete effectively both at home and abroad. Decisionmaking within the International Telecommunication Union (ITU) and other specialized UN agencies has also begun to be needlessly politicized due to block voting by lesser developed countries. The one-nation, one-vote premise upon which these multilateral international bodies ostensibly function admits to further politicization contrary to the interests of developed nations, especially the United States.

These specific national and international developments are more than isolated instances posing resolvable short-run problems for the United States. Collectively, they reflect the emergence of restrictive trends in the international telecommunications and information environment. Projected into the future, a gradual erosion of the United States position in the telecommunications, information flow, and associated high-technology markets may result, absent prompt remedial action.

The dispersal of responsibility and the lack of policy authority at the highest levels of our Government have prevented the United States from responding effectively and quickly to this escalating challenge to its defense, economic, and political interests. It has also adversely affected the ability of U.S. firms to function effectively abroad by signalling to foreign administrations that these key "sunrise, high-tech" industries are not valued sufficiently by the Nation's political leadership to warrant full collateral support of industry's efforts. It is important that remedial measures be initiated now. If we wait another ten years before we counter these adverse trends — after seven more critical international conferences and many more attacks on the free trade of goods, services, and ideas — it will be too late. The likely impact on U.S. defense capabilities, on employment and economic growth, and on freedom itself will be catastrophic.

The list of specific adverse events that buttresses these harsh conclusions is long. However, it is critical that policymakers in the Executive, as well as in Congress, become familiar with those of greatest significance. These events are set forth below in chronological order.

- 1970: Technology Targeting — Japan includes "knowledge intensive industries" in its national economic policy and planning, thus recognizing the importance of information products in shaping future Japanese economic development.

- 1972: Information Control -- Brazil establishes a Coordinating Commission on Data Processing Activities (CAPRE) to promote development of its indigenous telecommunications and information infrastructures and to control the flow of information in its economy. The resulting policies severely limit access to the Brazilian market and serve as models for other developing countries in promoting their own telecommunications and information sectors.

- 1973: Information Control — European countries commence enacting data protection laws designed to control international flows of personal data ostensibly to safeguard personal privacy. Some laws affect non-personal corporate data as well, however, thus potentially handicapping U.S.-based multinationals.

- 1973: ITU Politicization -- The politicization of the ITU, a process that began in 1965, continues with the strengthening of developing country voting blocs and the expulsion of Portugal and South Africa from the Torremolinos Plenipotentiary Conference. One-fourth of the Conference is consumed by heated debates over membership and other political topics, not radio frequency management issues.

- 1976: Trade Restriction — At its Sixth Plenary Session, the International Telegraph and Telephone Consultative Committee (CCITT) of the ITU adopts recommendations prohibiting resale and shared use of private lines.

- 1976: Trade Restriction — Canada denies tax deductions to Canadian businesses for advertisements aimed at Canadian viewers, but broadcast on foreign stations. A significant reduction in the revenues of U.S. border broadcast stations results.

- 1976: Information Control -- A variety of measures surface in UNESCO supposedly to reverse perceived North-South inequalities in the telecommunications and information fields. Included are proposals for a New World Information and Communication Order and a code of ethics and, indeed, the licensing of journalists.

- o 1978: Trade Restriction — Japan unilaterally imposes restrictive conditions on leased channel service sought by Tymshare, Inc., and Control Data Corporation. Each leased circuit must terminate at a single facility in the United States, thereby preventing these U.S. data processing companies from offering a full line of services in the Japanese market.

- o 1978: Technology Targeting — The Nora-Minc Report to the President of France expresses the need for a national strategy to control the impact of "telematics" on society, develop indigenous computer and telecommunications capabilities, and respond to the "renewal of the IBM challenge."

- o 1979: Information Control — A Canadian government study, the Clyne Report, recommends that "the government should act immediately to regulate transborder data flows to ensure that we do not lose control of information vital to the maintenance of national sovereignty."

- o 1979: Technology Targeting — The Japanese Ministry of International Trade and Industry (MITI) issues "MITI Vision for the 1980s." This policy report recommends targeting the computer and data processing industries as crucial to Japan's long-term economic progress.

- o 1979: ITU Politicization — The World Administrative Radio Conference fails to resolve national differences regarding the use of the geostationary satellite orbit. Passed is a resolution sponsored by a block of lesser developed countries to convene a conference to "guarantee in practice for all countries equitable access to the geostationary satellite orbit and the frequency bands allocated to the space services."

- o 1979: Trade Restriction — The European Commission (EC) issues a report recommending a community-wide strategy to develop telecommunications and information markets, to improve European capabilities in information services, and harmonize standards. EURONET, a community-wide data communications network, reflects the goal of providing purely European services in the EC market.

- o 1979: Trade Restriction -- European countries refuse to include the Post, Telegraph, and Telephone (PTT) administrations as "government agencies," thus subject to the GATT goverment procurement code.

- o 1980: Trade Restriction — The Canadian Banking Act is enacted and prevents transactions being processed outside Canada unless processing is done domestically as well. Prior approval before financial data can be sent out of the country is also required.

- o 1980: Information Control -- The MacBride Commission Report on international communications is transmitted to the UNESCO Director General. Its findings stridently support Third World demands for "more justice, more equity, more reciprocity in information exchange, less dependence in communication flows, less downwards diffusion of message, more self-reliance and cultural identity, more benefits for all mankind," but suggest severe restraints on Western news media.

- 1980: Information Control — The Council of Europe adopts a treaty concerning protection of individual privacy to be legally binding when ratified by five member nations. Once ratified, this treaty could seriously restrict flows of personal data to non-member countries.

- 1980: Trade Restriction -- A French report by Alain Madec asserts transborder data flows reinforce the economic strength of multinational companies and "even more than trade in products, mean the decay of the state." The report presents a scheme for analyzing information as a commodity, which may serve as the basis for imposing customs duties and value-added taxes on transborder data flows.

- 1981: Technology Targeting — MITI sponsors the Fifth Generation Computer Conference and outlines a ten-year government-industry R&D program by which the Japanese hope to leapfrog the U.S. computer industry.

- 1981: Trade Restriction — The United States Trade Representative compiles an inventory of over 100 non-tariff trade barriers posing current and potential problems for the telecommunications, data processing, and information services areas.

- 1982: ITU Politicization — Amendments to the ITU Convention at the Nairobi Plenipotentiary Conference expand the purposes of the ITU to include provision of technical assistance to developing countries. Also changed are certain election procedures for directors of the International Consultative Committees who will no longer be elected by their technical peers in plenary sessions, but by all participants in plenipotentiary conferences.

- 1982: Trade Restriction — West German Bundespost regulations go into effect which condition private leased line access to international lines on the local processing of data before international transmission. These regulations serve the dual purpose of protecting the domestic data processing industry and increasing PTT revenues from new volume sensitive services.

This list is not only a litany of concerted actions taken by other nations. It is also an indictment of the lack of U.S. policy coordination in the face of ever-increasing economic and political challenges. The report that follow sets out in detail how we reached the adverse situation in which we find ourselves, what our policy and organizational options are, and which of these options offer the best chance of enhancing U.S. and free world interests.

There is great international strength in U.S. ideas, technology, and free enterprise. There is great weakness and danger in complacency and indecision.

Introduction

OBJECTIVES OF THE REPORT

The primary objective of this report is to provide a comprehensive delineation of the goals, policies, strategies, and principal issues in the international telecommunications and information field in order to improve the formulation and execution of Government policy. While this report constitutes an important step, the only effective way to ensure consistent and effective policy is for private enterprise, Congress, and the Executive branch to assert a level of commitment to the field commensurate with its significance for U.S. interests and to see that a proper organizational scheme is established with clear responsibility for maintaining high performance in policy formulation and implementation on an ongoing basis.

Chapter One

INTERNATIONAL TRENDS IN TELECOMMUNICATIONS AND INFORMATION

A number of international telecommunications and information developments have serious implications for U.S. interests. These sectors are crucial for the United States and a strong presence in international telecommunications and information markets is essential to our economic vitality. The long-held leadership position of the United States has been challenged by other countries, however, which also consider these sectors of vital national importance.

This section highlights the significance of these high-technology, "sunrise" sectors for the U.S. economy and the growing foreign competition we confront. It discusses the twin underlying trends which pose major problems for U.S. policy:

(1) the growing prevalence of trade barriers and other protectionist policies once reserved for traditional labor-intensive, "smokestack" industries; and,

(2) the increased politicization of the issues in international forums.

Meeting these challenges calls for concerted Government action; instead, basic deficiencies in the coordination of U.S. policy and the level of attention it receives have been exposed.

Significance for U.S. Interests

Relative to other parts of our economy, the telecommunications and information sectors have experienced rapid, indeed, exponential growth in the past two decades and become driving forces of change in contemporary society. Technological advances, for example, have triggered the shift from an industrial to a service-oriented economy. Services-related industries are information intensive and thus depend heavily on advanced communications and computer systems to provide necessary access to and transfer of information. The strong link between the information and telecommunications sectors extends beyond the domestic sphere, however, to international markets as well.

As world leaders in information and telecommunications technologies, U.S. firms have employed their technological talents to serve foreign markets and consumers. Exports of information and telecommunications goods and services have made vital contributions to the U.S. balance of trade. International telecommunications and information flows are also crucial to the efficient operation of U.S.-based multinational firms. Virtually all firms with overseas operations rely heavily on international information flows to conduct business. Reliable and cost-effective access to the telecommunication facilities and services in the countries where subsidiaries are located is thus necessary. Net earnings from such U.S. overseas holdings amounted to $24.1 billion in 1981. These firms also accounted for a significant proportion of U.S. revenues from services exports.

The information and telecommunications sectors are not only important as growth sectors themselves. They also function as supporting factors in the growth of other industries -- and constitute major contributors to restoring the strength and productivity of the U.S. economy. Until the 1970s, U.S. firms dominated international markets in high technology goods and services. Since then, however, foreign firms have made inroads in several key areas, including semiconductors, robotics, microcomputers, lasers, and satellite communications. Japan's early market lead in 64K random access memory (RAM) chips is just one example of such increasing foreign competition.

Protectionism and Other Anticompetitive Practices

An increasing number of countries have targeted the information and telecommunications sectors for special government support and protection against foreign competition, in recognition of their critical role in future development. As a result, a variety of anticompetitive measures are currently in place including:

- denying or restricting access by U.S. firms to foreign telecommunications and information equipment and services markets;
- devising technical interface and equipment standards which needlessly preclude or hamper use of foreign-owned equipment in connection with domestic public telecommunications networks;
- extending concessionary export financing for domestic firms;
- imposing local equipment purchase requirements or local content requirements;
- providing extraordinary tax incentives, direct subsidies, or low cost loans for research and development to local firms;
- restrictive government procurement; and,
- imposing higher rates for private-line services for the purpose of excluding U.S. competition.

The current global recession, with its attendant pressures on national economic policies, has only aggravated the tendency to erect protectionist barriers in the telecommunications and information sectors. In addition to outright protectionism, the United States faces anticompetitive policies growing out of the regulatory traditions of telecommunications markets. In most other countries, telecommunications facilities are owned and services are controlled by the government, and the interests of some government-owned post, telegraph, and telephone (PTT) entities have hindered the development of open markets. Some nations, including Britain, Australia, and Japan, have voiced an intention to move rapidly toward more freely competitive conditions. Other PTTs, however, seem intent on maintaining their traditional monopolies. While U.S. experience suggests competition can expand overall demand for telecommunications services, some abroad are concerned lest their present revenues be eroded as a result of foreign competition.

Finally, many countries have become more concerned about maintaining sovereignty over the transmission of information across their borders, ostensibly for national security and socio-cultural reasons. U.S. firms operating overseas, for example, are concerned about the possible adverse impact of recent European data protection laws. Designed in theory to protect individual privacy in the face of sophisticated data processing techniques, these laws either expressly prohibit or authorize restrictions on the export of personal data. A potential consequence of these broadly phrased data protection laws, however, might be needless interruptions or restrictions on international transfers of non-personal data.

The serious danger exists that the cumulative effect of these protectionist and other barriers to U.S. competition will be to curtail trade and direct investment in telecommunications and information equipment and services, reduce the flexibility with which users apply telecommunications and information technologies, and lead to further restrictions on the free flow of information. An immediate problem is the disincentives to foreign investment created by uncertainty about future trade barriers.

Politicization of International Forums

As the international institution directly responsible for managing the technical aspects of international telecommunications, the International Telecommunication Union (ITU) has become a focus for Third World efforts not only to gain needed assistance in the telecommunications field, but also to further other, often unrelated, political ends. What was once chiefly a forum for the quiet exchange of engineering views and judgments has become embroiled in many of the same controversies affecting other international forums. The attempted expulsion of Israel at the 1982 Plenipotentiary Conference in Nairobi is the most recent example. At Nairobi, there was also a concerted, and successful, effort by Third World countries to expand the role of the ITU to include the provision of technical assistance to developing countries.

Other international organizations have placed international information and telecommunications issues at the forefront of their agendas. These include the Organization for Economic Cooperation and Development (OECD), the Council of Europe, the United Nations Educational, Scientific, and Cultural Organization (UNESCO), the United Nations Center on Transnational Corporations (UNCTC), and the Intergovernmental Bureau for Informatics (IBI). (Profiles of these and other organizations are set forth in an Appendix to this report.)

The importance accorded the telecommunications and information sectors, and the diversity of national interests and levels of development reflected in these international organizations, have elevated international discussion to a highly sensitive, political level of attention.

Activities in the special agencies of the U.N. clearly reflect Third World interests. Among developing countries, there is widespread sentiment that the existing international framework does not serve to lessen global inequalities in telecommunications and information resources and capabilities. A majority of Third World governments have focused their efforts on redressing the North-South imbalance in these critical sectors by collective actions.

The call for a New World Information Order through UNESCO is just one example of such collective action. Agenda-setting in the IBI and the UNCTC is another; both institutions are concerned primarily with the problems of developing countries.

In 1980, the OECD adopted a set of voluntary guidelines for corporations and governments to follow with regard to the protection of individual privacy. In contrast to these voluntary guidelines is a proposed treaty, covering the same subject, for the member countries of the Council of Europe. Pending ratification, this treaty could seriously restrict flows of personal data to non-member countries.

The increasing politicization of the international forums in which telecommunications and information issues are debated poses problems for the orderly management, by consensus, of those issues for the benefit of all nations. Most international organizations operate on the one-nation, one-vote principle. It is thus likely the United States will continue to experience difficulties in respect of decisions and actions taken at the international level. The growing number of international organizations involved with telecommunications and information issues also increases the likelihood any international "rules of the road" developed governing the activities of nations and private entities will impinge upon U.S. interests.

U.S. Government Response

As indicated elsewhere in this report, the U.S. Government has undertaken false starts in seeking to prepare itself to respond to these situations. Policy has evolved in piecemeal fashion. Problems have been aggravated by inadequate high level attention and insufficient coordination among the diverse departments and agencies involved. The net result too often has been confusion, needless jurisdictional disputes, and consequent lack of adequate preparation -- all of which place the United States at a serious disadvantage. Reexamination of our strategy and Government organization for pursuing U.S. telecommunications and information goals is required. The interests at stake demand high level attention and serious political commitment to ensure that U.S. interests are not compromised by default.

Chapter Two

GOALS, POLICIES, AND STRATEGIES

Our basic long-range telecommunications and information policy goals are those which, if achieved, will provide a stable national and international environment commensurate with our basic principles of national existence. These policy goals are defined both directly by Constitutional, Legislative, and Executive parameters and indirectly by obvious national interest considerations. Our principal policy goals include assuring:

- the free flow of information worldwide, subject only to the most compelling national security and personal privacy limitations.
- the necessary growth of the national security, public service, and commercial interests of the United States occurs in a manner commensurate with our leadership role in the world.
- that information flow to developing nations contributes fully to the elimination of hunger, poverty, disease, and ignorance and facilitates their sound economic development.
- there is a free and competitive marketplace for telecommunications and information services equipment and facilities.
- there are efficient non-political international organizations for the development, management, expansion, and non-discriminatory access to international telecommunications facilities and networks.
- that human well-being and understanding grow as rapidly as possible through international telecommunications services.

These goals are fundamentally compatible and can be pursued individually or together. The national interest may require, however, that some goals be accorded priority over others; and these priorities may also shift from time to time. Hence, there will always be a need to balance these various goals in formulating and implementing international telecommunications and information policy.

TWO PRINCIPLES FOR POLICY

In working to attain the goals enumerated above, U.S. telecommunications and information policy has been moving with reasonable consistency and to a greater degree than any other nation toward reliance on two broad principles: free flow of information and free competitive market enterprise. Thus, in general, individual policy decisions formulated to achieve long-range goals in international telecommunications and information reflect efforts to:

- enhance the free (without restriction or control) flow of information across national borders, with limited exceptions condoned only for the most compelling reasons; and
- promote an international environment for the provision of telecommunications and information facilities, services, and equipment — and for the production and dissemination of information itself — in which maximum reliance is placed on free enterprise, open and competitive markets, and free trade and investment with minimum direct government involvement or regulation.

These principles currently guide U.S. policy in many parts of the international telecommunications and information sector. With respect to international facilities and services, our basic policy was succinctly stated in a recent report by the Senate Committee on Commerce, Science, and Transportation:

> The policy of the United States is to rely wherever and whenever possible on marketplace competition and the private sector to provide international telecommunications services, and to reduce or eliminate unnecessary regulation. This is based upon the Committee's belief that competition enhances technological innovation, efficiency, and provision of services to the public at reasonable rates. When it is necessary to regulate international telecommunications services, it must be the absolute minimum necessary to achieve the purposes of the act.[1]

Despite a tradition of heavy Government regulation of the international telecommunications industry, a policy of fostering maximum feasible competition has been adopted and implemented by the Federal Communications Commission (FCC) as the basis for regulation of international facilities and services.[2]

U.S. policy emphasizes fundamental marketplace principles in other areas as well. With respect to international trade in equipment and services, the United States is the foremost advocate of liberalizing and reducing barriers among the countries of the world. Similarly, U.S. policy for research and development in this sector relies heavily on technological innovation achieved primarily through private initiatives.

Finally, with respect to mass media and other issues of information policy, the United States persistently has called for worldwide recognition of the principle of free flow of information with minimum government interference. In addition to its economic benefits, free flow of information in the "marketplace of ideas" serves to promote cultural development and to strengthen political liberty and effective self-government.

Reliance on the marketplace and free flows of information establishes basic guidance for formulating policy. In some cases, however, achieving U.S. goals requires limited Government intervention. Where market structure necessitates, some Government oversight (e.g., natural monopoly), where close cooperation between the United States and other sovereign nations is vital (e.g., for the allocation of radio spectrum), or where the unfettered marketplace will not necessarily achieve important ends (e.g., in matters of national security or foreign policy), governments must intervene in telecommunications and information activities. Nevertheless, there is consensus within the United States that reliance on market principles is generally consistent with our international telecommunications and information objectives, and that, when government intervention is required, it should be structured to minimize interference with economic efficiency, competition, and the free flow of information.

PROBLEMS IN APPLYING THE PRINCIPLES

Two significant problems are encountered when policy is based on market principles and free flow of information. First, there is not yet a consensus among other countries that these principles should apply to telecommunications and information services, and second, the concept of free flow of information admits to a number of interpretations. Each of these points merits elaboration.

Foreign Resistance to Market Principles and Free Flow

Throughout the international telecommunications and information arena, the United States has encountered resistance by other countries to the application of marketplace and free flow principles. In the International Telecommunication Union (ITU) and other forums where countries collaborate on the planning of facilities and services, it has been difficult to obtain agreement on competitive, efficiency-enhancing policies. Most countries continue to follow a monopolistic approach to telecommunications service. Some major countries commendably have begun to move toward reduced government control. The majority, however, remain unconvinced of the benefits of competition. As a result, the foreign telecommunications administrations (PTTs) have refused to conclude operating agreements with new U.S. entrants in the international services market, or have subjected them to severe regulation. There is also some danger that some PTTs will seek to use their monopoly power unfairly to exploit the increasingly competitive environment of the United States.

Second, in the area of radio spectrum and satellite orbit management, the developing countries have increasingly opposed the allocation of frequencies and orbital positions on the basis of economic efficiency. This opposition reflects the growing politicization of the ITU and a commensurate reduction of its effectiveness in solving technical problems.

Third, international trade in equipment and services is increasingly disrupted by industry-targeting policies of other governments, including the use of subsidized export financing and the erection of protectionist non-tariff trade barriers. There is significant concern in this country that, without a reduction in such practices, our continued adherence to a policy of open markets and minimal government intervention will ultimately harm U.S. interests.

Finally, in the area of mass media and information policy, UN organizations have drafted proposed "codes of conduct" in support of restrictive policies. We believe these codes fail to strike a reasonable balance between legitimate concerns over sovereignty and the fundamental doctrine of free flow of information advocated by the United States. In addition, there is a trend toward greater government control of "transborder data flows" of both commercial and personal information circulating among computers located in different countries.

Despite the disparity in acceptance of marketplace and free flow principles between the United States and other countries, NTIA believes that we should continue to adhere to these principles as guideposts for U.S. telecommunications and information policy in coming years. The principles themselves should not be abandoned, although the strategy for their implementation requires improvement.

Free Flow of Information

A basic difficulty with grounding U.S. policy on the concept of free flow of information is the variety of interpretations of this phrase. This variety can be attributed, in part, to the peculiar characteristics of information. Under certain conditions, information assumes the attributes of an intangible "commodity," with market-determined value — a product that commands a price from its consumers. Under other circumstances, it constitutes an "intermediate resource" applied at various stages in the process of producing other goods and services. Under still other conditions, it conveys fundamental beliefs or rudimentary ideas to which economic value cannot be objectively assigned and for which regulations to achieve economic goals may clash with basic rights of free thought and expression.

Among the meanings that might be attributed to "free flow" are:

- o An extension of the First Amendment prohibition against laws "abridging the freedom of speech, or of the press," and in the sense expressed by Article 19 of the Universal Declaration of Human Rights;
- o The absence of impediments imposed <u>inadvertently</u> as a consequence of regulations not directed at information flows per se;
- o The absence of laws or regulations that <u>intentionally</u> impose restrictive conditions on the location of data-processing facilities, or on the transmission beyond borders of certain kinds of information (the motivations may be purely economic and not political, and content may be of no particular concern to the regulators);
- o The absence of governmental attempts to require disclosure—for economic, social or political reasons -- of the content of information being processed or transmitted;
- o The availability of information without direct cost to the recipient.

United States policy in international telecommunications and information should continue to be grounded on the basic principle of free flow of information. Clarification of how this principle applies in various circumstances, however, can be made.

- o Free flow of information means unrestricted flow of information. It does not refer to matters concerning the allocation among recipients of the costs of production and distribution.
- o With regard to freedom of speech and press, and international flows of information via the print and broadcast media, policy will continue to provide unequivocal support for free flows.
- o With regard to information as an economic commodity, policymakers should recognize that it is rarely necessary to regulate information itself in order to achieve legitimate economic objectives. Ordinarily, the costs of such regulation outweigh any economic or social benefits.

The U. S. position on this last matter, consistent with the long-range goals of promoting telecommunications and information technology as a contributor to efficient resource utilization, would be to oppose strongly any actions interfering with the ability of producers and users to make optimum use of information as a productive resource. This will lead to a more efficient utilization of resources. It will also lead to greater revenues for both private entities and, ultimately, for taxing authorities.

Adopting free flow of information as a basic principle of policy has not meant, nor will it in the future mean, that the United States does not concede the need for exceptions for certain reasons. The requirements of maintaining national security is one example. Here, too, however, any impediments should be held to an absolute minimum and imposed only when doing so will clearly and efficiently achieve the desired objective.

In summary, although the "free flow of information" principle is subject to several possible interpretations, it stands essentially for the least possible control of information by governments for any reason. How this general position will apply in individual circumstances -- and the many "new" circumstances created by advancing technology — is a matter for on-going consideration.

POLICY IN SPECIFIC AREAS

The report covers six major areas relevant to international telecommunications and information. These are: (1) research and development, (2) facilities and networks, (3) international telecommunications services, (4) trade in merchandise and services, (5) information, and (6) national security. Each of these areas are discussed in separate portions of this report. These are the principal findings and recommendations:

Research and Development

An important objective of U.S. research and development policy is maintaining scientific and technological leadership in the telecommunications and information industries. Traditionally, the United States has relied heavily on private initiative to assure adequate innovation. In the future, private initiative must remain the primary source of technological development. There are two factors, however, that create a need for Government attention to reinforce and complement the activities industry has undertaken: (1) the high cost, high risk nature of R&D in telecommunications and computers, and (2) the "targeting" policies of other governments that have accelerated the rate of technological advance of competitors. In recognition of these factors, the following general policies are appropriate for supporting the overall goals listed earlier:

- o Heighten Federal support for R&D through
  - increased direct Federal funding of basic research
  - improved tax credits
  - liberalized patent rights for corporations;

- o Establish a mechanism to obtain outside advice for Federal R&D activities;
- o Continue aggressive support for joint research activities among government, universities, and business;
- o Permit greater cooperation among U.S. business through joint R&D ventures;
- o Provide Federal assistance in support of R&D activities in small innovative firms;
- o Improve data gathering on the position of the U.S. relative to other countries in technological standing.

Many of these points are endorsements of Federal actions already underway. This Administration's basic economic policies, however, have also had an affirmative impact. By sharply reducing inflation and restoring needed stability and predictability to Government decisionmaking generally, the Reagan Administration has gone far toward providing a commercial environment conducive to long-term product development and basic research rather than short-term profit seeking as engendered by previous eclectic economic policies.

Facilities and Networks

With respect to facilities and networks, five issues are discussed: (1) allocation of spectrum resources; (2) allocation of satellite orbital resources; (3) facilities planning and authorization; (4) Comsat and Intelsat issues; and (5) integrated services digital networks (ISDN).

Policy in the provision of international facilities and networks in general should track closely fundamental principles favoring market competition. Emphasis should be placed on the efficient use of scarce resources, flexible planning responsive to technological changes, alleviation of bottlenecks in facilities, and dissolution of unnecessary or unfair monopoly advantages. Future facilities, particularly ISDN, should be carefully designed to accommodate the needs of users and to maximize the likelihood of competition which will benefit those users. Specifically, in:

- o Allocation of Spectrum and Geostationary Orbit Position
  - We should adhere to current policies that favor allocation on the basis of efficiency and established need, while assuring that the needs of future users will be effectively met.
- o Facilities Planning
  - We should seek new ways to facilitate consultation with other

administrations while reducing or eliminating unnecessary regulatory delays.

- Comsat/Intelsat
    - The United States should continue support for the Intelsat system.
    - We should promote unrestricted ownership of earth stations in the United States.
- Integrated Services Digital Network
    - The Government in close cooperation with private sector users and service providers should develop a more formal policy regarding the evolution of ISDN to assure greater U.S. influence in the international process of developing network configuration and standards.

International Telecommunications Services

It is the policy of the United States to place maximum reliance on marketplace competition wherever possible in the provision of international telecommunications services and to reduce or eliminate unnecessary regulation. The United States, however, must retain sufficient Government oversight authority to assure the success of its competitive policies and to safeguard vital U.S. interests in national security, foreign policy, trade, and technological leadership. With respect to international services, the United States should:

- assure that private line service remains a network option for international telecommunications and that ISDN provides comparable or superior choices;
- support the efforts of new U.S. service providers to obtain international operating agreements and prevent any carrier from entering exclusive contracts with foreign administrations;
- continue a uniform accounting and settlement policy for switched services;
- retain sufficient oversight and take appropriate steps to protect U.S. carriers and service providers from unfair competition by foreign government-affiliated entities;
- ensure nondiscriminatory interconnection and fair competition in international voice service; and,
- maintain a strong U.S. role in the CCITT.

Trade in Equipment and Services

The telecommunications and information industries require greater attention in overall U.S. trade policy. Telecommunications and information equipment and services in the past decade have assumed much greater importance both in the U.S. economy as a whole and in the U.S. balance of trade. Although U.S. trade policy in this sector (as in others) adheres firmly to the principles of free trade and open competition, our ability to continue making trade policy on this basis is being tested by the proliferation of trade barriers throughout the world.

In the telecommunications sector, barriers to trade and investment in equipment and to the international supply and use of services have long existed. They pose especially pressing problems today, as the U.S. industry is deregulated and focuses increasingly on serving both domestic and foreign customers. In the information sector — computers and data processing — U.S. industry's leading position is being challenged by the development of hardware manufacturing and data processing capabilities in other countries, often aided by concerted industry-targeting policies.

These problems have grown in importance at the same time the overall world economic picture has darkened. Protectionist sentiments are increasingly prevalent. It is thus crucial that the telecommunications and information industries be given greater attention on the U.S. trade policy agenda. In particular, the United States should:

- place a high priority on the reduction of non-tariff trade barriers affecting the telecommunications and information industries through vigorous multilateral and bilateral negotiations in the GATT and elsewhere, but without insisting on rigid sectoral reciprocity;

- take appropriate actions — including, if necessary, the amendment of U.S. trade laws — to protect U.S. telecommunications and information industries from unfair industry-targeting practices and other anticompetitive policies of other countries;

- augment the U.S. export promotion effort in the telecommunications and information sectors by identifying and reducing or eliminating U.S. barriers to exports; and

- assure the integration of telecommunications and information services into the overall U.S. trade effort, by identifying the barriers encountered by U. S. suppliers and users of such services abroad and vigorously seeking their reduction.

Information

The basic recommendations regarding information have already been presented above in the chapter on "Free Flow of Information." On the specific subjects discussed, the following recommendations apply:

- Press Freedom. U.S. policy will continue uncompromised support for a free press and free international flows of information.

- o Communications and Development. Private initiative is providing the expertise and guidance needed to develop the telecommunications and information sectors of developing countries. Greater efforts by the private sector will be mutually beneficial. In addition, Government agencies responsible for foreign aid should review the priority accorded assistance for communications development.

- o Direct Broadcast by Satellite (DBS). U.S. preparations for the 1983 Region 2 Regional Administrative Radio Conference on DBS planning are well underway, but the recent UN vote on DBS should encourage increased concentration on the political aspects of our groundwork. Several U.S. Government agencies involved in international broadcasting have been investigating the potential of DBS. This activity should continue with appropriate safeguards and notification bounds of existing arrangements to protect the interests of U.S. businesses in this area.

- o Privacy Protection. U.S. policy should continue to recognize the need for personal privacy protection, and support efforts of individual countries to implement safeguards according to their own legal traditions.

- o Valuation and Taxation of Information. Consistent with the objective of promoting the role of telecommunications and information technology, the United States strongly opposes any actions that would interfere with the ability of producers and users to make optimum use of information as a productive resource.

- o Encryption. Efforts should be undertaken to formulate a clear U.S. policy on encryption that will accommodate both the legitimate concerns raised by national security and the needs of users of international facilities and networks.

- o Intellectual Property Rights. The United States should maintain close contact and cooperation with other countries to ensure the development of mutually acceptable forms of protection for property rights for new forms of intellectual property and continue within the bounds of existing arrangements to protect the interests of U.S. businesses in this area. In this regard, the Government should ratify promptly the Brussels Convention concerning unauthorized commercial reception and use of copyrighted material transmitted by satellite.

National Security

National security concerns bear on each of the areas discussed in this report. Ensuring national security is a fundamental goal, and telecommunications and information are crucial factors in achieving that goal. Included within the scope of national security are not only national defense and related military concerns, but also the conduct of foreign policy, the economic strength of the nation, and emergency preparedness.

As telecommunications and information technology evolves and the field becomes more competitive, steps will have to be taken to assure the specific needs of the national security community are satisfied. These include ensuring the availability of reliable and economical telecommunications networks, the security of messages transmitted, and adequate procedures for restoring networks in case of national or international emergency. In addition, a more effective means of including the defense community in the process of formulating and executing international telecommunications and information policy is vital.

Chapter Four

ORGANIZATION AND STRUCTURE OF THE U.S. GOVERNMENT

BACKGROUND

The ability of the U.S. Government as currently organized and structured to establish and achieve long-term international telecommunications and information policy goals and objectives is being severely questioned today. Spokespersons from private industry, members of Congress, Executive branch policymakers, even representatives of foreign governments, are highly critical of U.S. performance to date. They are especially concerned about Government capability to protect vital national interests in future negotiations, conferences, regulatory proceedings, and legislative initiatives.

More than two dozen departments and agencies of the Federal Government are involved in the development, implementation, and operation of U.S. international telecommunications and information policy.[1] For some it is a primary mission; for others an occasional sidelight. But each has different expertise, tools, and available forums with which to seek its goals, and each brings a different perspective, if not constituency, to an issue. While the argument is made that diversity and cross-fertilization yield strength, a review of international telecommunications policymaking in the U.S. suggests that they also breed confusion, conflicts, jurisdictional disputes, lack of coordination, and lack of adequate preparation. These concerns are not new.

During its 1946 investigation of international communications problems, the Senate Committee on Interstate Commerce sought "to obtain a well-rounded and over-all viewpoint from the Government agencies. . . ."[2] The Committee's Interim Report noted, however, that:

> During the course of the hearings it became obvious that the affected Government departments were not of one mind with respect to the policy that should be laid down by this Government to govern American international communications . . . . The Executive had constituted an interdepartmental committee to study the problem and make recommendations . . . (but) this interdepartmental committee had failed to reach complete agreement.[3]

The Interim Report continued:

> The committee regards with favorable anticipation the recent creation of another interdepartmental committee and trusts that this agency will be able to shortly recommend to the President a unified departmental view.[4]

In 1951 President Truman's Communications Policy Board considered, among other questions, how the U.S. Government could "strengthen its organization to cope" with the major domestic and international telecommunications issues facing the Nation.[5] In what sounds like contemporary criticism, the Board reported that:

> In our efforts to discover the current state of Government telecommunications policy as preliminary to recommending needed steps toward a total national communications policy, we once more encountered dispersion, confusion, gaps, and deficiencies in the product and performance of those agencies charged with telecommunications policy responsibilities.[6]

In anticipation of developing national policy for dealing with other nations in seeking international telecommunications agreements, the Board found that:

> Just as the United States has no clear policy for apportioning its own share of spectrum space, so it has lacked satisfactory means of determining policy as a basis for negotiations with other nations.[7]

As for maintaining a sound private telecommunications industry, the Board found "there has been no long-range study of the question, no long-range planning. No agency of Government is in a position to take a comprehensive view of this problem."[8] Similarly, a 1968 Presidential Task Force on Communications Policy reported:

> Traditionally, government has viewed telecommunications primarily as a mission-support function, rather than a focus for public policy. The result has been that policy has evolved as a patchwork of limited, largely ad hoc responses to specific issues, rather than a cohesive framework for planning. Government organization for the formulation and implementation of communications policies reflects this evolution. . . . The patchwork nature of the present structure is not conducive to optimum performance of the telecommunications activities and requirements of the Federal Government.[9]

From these hearings, reports, and studies of the Forties, Fifties, and Sixties, to the 1982 report of the Senate Commerce Committee on S. 2469 ("The International Telecommunications Act of 1982"), the 1980 report of the House Government Operations Committee (<u>International Information Flow: Forging a New Framework</u>), and the Office of Technology Assessment study (<u>Radio Frequency Use and Management, Impacts from the World Administrative Radio Conference of 1979</u>), the persistent lack of an effective Government organization to develop, adopt, and implement U.S. international telecommunications policy has been widely discussed.

To recognize that these organizational problems are long-standing and difficult to resolve, however, does not diminish the necessity for prompt attention and comprehensive efforts to establish a more effective government structure to deal with them. Telecommunications and information have become increasingly vital components of our national security, international trade and economic well-being, and critical to our relations with allies and other countries alike. At one time, the United States might have been able to afford the direct and opportunity costs implicit in the present disordered and dispersed telecommunications policymaking structure. American industry, too, may not have required in its dealings with other nations the collateral support which the fact or perception of organized high-level Government can provide. Whatever else may be true, however, we have today reached a point where the persistent inability of Government to get its own telecommunications policy "house" in order threatens severely to affect the future efficient development of key high-tech, "sunrise" industries upon whose effectiveness so much of our economy, security, and national life stands to depend.

PRIMARY PLAYERS

Among the major participants in the formulation of U.S. international telecommunications and information policy are the National Telecommunications and Information Administration of the Department of Commerce (NTIA), the Department of State, the Federal Communications Commission (FCC), and the U.S. Trade Representative (USTR). The role of each is discussed briefly below. Also described are the Interagency Group on International Communications and Information Policy, which was established by the State Department, and the Cabinet Council on Commerce and Trade, established by President Reagan. Each can be viewed as a potential source of the effective interagency coordination which has been lacking to date, and each will be discussed further.

National Telecommunications and Information Administration Department of Commerce

NTIA, a part of the Department of Commerce, has been delegated broad authority in both domestic and international telecommunications and information policymaking and operations. Among NTIA's responsibilities which are significant for international policymaking are to:

1. serve as the President's principal adviser on telecommunications policies pertaining to the Nation's economic and technological advancement and to the regulation of the telecommunications industry;

2. develop and set forth, in coordination with the Secretary of State and other interested agencies, plans, policies, and programs which relate to international telecommunications issues, conferences, and negotiations. The Secretary of Commerce shall coordinate economic, technical, operational and related preparations for United States participation in international telecommunications conferences and negotiations. The Secretary shall provide advice and assistance to the Secretary of State on international telecommunications policies to strengthen the position and serve the best interests of the United States, in support of the Secretary of State's responsibility for the conduct of foreign affairs;

3. provide for the coordination of the telecomunications activities of the Executive Branch, and . . .assist in the formulation of policies and standards for those activities, including but not limited to considerations of interoperability, privacy, security, spectrum use and emergency readiness;

4. assign frequencies to radio stations or to classes of radio stations belonging to and operated by the United States;[10]

5. develop and set forth telecommunications policies pertaining to the Nation's economic and technological advancement and to the regulation of the telecommunications industry;

6. ensure that the Executive Branch views on telecommunications matters are effectively presented to the Federal Communications Commission and, in coordination with the Director of the Office of Management and Budget, to the Congress;

7. conduct studies and make recommendations concerning the impact of the convergence of computer and communications technology.[11] [Emphasis added.][11]

Functions relating to the international communications satellite system, vested in the President by the Communications Satellite Act of 1962,[12] have also been delegated to NTIA.

NTIA was formed in the spring of 1978 as successor to both the Office of Telecommunications Policy in the Executive Office of the President and the Office of Telecommunications in the Commerce Department. Presidential Reorganization Plan No. 1 of 1977[13], and Executive Order 12046[14] transferred to the Secretary of Commerce significant policy and administrative functions in both the domestic and international telecommunications and information areas. They also established the position of Assistant Secretary of Commerce for Communications and Information. The Secretary of Commerce then created NTIA, prescribing the scope of its

authority and the functions of the Assistant Secretary.

Rather than providing a clear demarcation of international telecommunications policymaking responsibilities and authority, however, Executive Order 12046, Congressional authorizations and appropriations, and underlying statutes do not clearly establish in many instances the responsibilities of NTIA, the Department of State, the FCC, USTR, the Department of Defense, and other agencies.[15]

Department of State

The Secretary of State serves as principal foreign policy adviser to the President, and is responsible for the overall direction, coordination, and supervision of U.S. foreign relations. The conduct of international telecommunications and information policy has been treated as a minor subset of that function. Executive Order 12046 in 1978 appeared to reaffirm this view, in its direct grants of telecommunications authority to the Secretary of State:

> With respect to telecommunications, the Secretary of State shall exercise primary authority for the conduct of foreign policy, including the determination of United States positions and the conduct of United States participation in negotiations with foreign governments and international bodies. In exercising this responsibility the Secretary of State shall coordinate with other agencies as appropriate, and in particular, shall give full consideration to the Federal Communications Commission's regulatory and policy responsibility in this area.
>
> . . .
>
> (The Secretary of State shall) [e] xercise the supervision provided for in Section 201(a)(4) of the Communications Satellite Act of 1962, as amended (47 U.S.C. 721(a)(4)); be responsible, although the Secretary of Commerce is the chief point of liaison, for instructing the Communications Satellite Corporation in its role as the designated United States representative to the International Telecommunications Satellite Organization; and direct the foreign relations of the United States with respect to actions under the Communications Satellite Act of 1962, as amended. [Emphasis added.][16]

Executive Order 12046 reaffirmed this view by its mandate to the Secretary of Commerce as well:

> The Secretary (of Commerce) shall provide advice and assistance to the Secretary of State on international telecommunications policies to strengthen the position and serve the best interests of the United States, in support of the Secretary of State's responsibility for the conduct of foreign affairs.[17]

The Department of State heads or names the head of U.S. delegations to international telecommunications conferences and negotiations; delegation members are selected by the State Department after consultation with involved Federal agencies and consideration of the input of the private sector. At times other agencies have been designated by the State Department to represent the United States at specific meetings or for special purposes.

In light of growing criticism of the international telecommunications policymaking performance of the Executive Branch, an Interagency Group on International Communications and Information Policy, chaired by the Under Secretary of State for Security Assistance, Science, and Technology, was created. It was designed as "a senior-level group which ensures coordinated development of policy by the interested departments and agencies of the Executive Branch which includes participation of the Federal Communications Commission."[18] Those who attend its irregularly scheduled meetings include representatives of the Departments of State, Commerce, and Defense, USTR, OMB, OSTP, NSC, NASA, the Board for International Broadcasting, USIA, AID, CIA, and the FCC.

Federal Communications Commission

The FCC was created by the Communications Act of 1934[19] as an independent regulatory agency responsible directly to the Congress. Although its seven (soon to be five)[20] Commissioners are appointed by the President, once confirmed by the Senate for seven-year terms, they do not serve at the pleasure of the President. The FCC thus is not a part of the Executive branch, nor is it necessarily bound to Administration policies. The FCC was created:

> For the purpose of regulating interstate and foreign commerce in communication by wire and radio so as to make available, so far as possible, to all the people of the United States a rapid, efficient, nation-wide, and world-wide wire and radio communication service with adequate facilities at reasonable charges, for the purpose of the National defense, for the purpose of promoting safety of life and property through the use of wire and radio communication, and for the purpose of securing a more effective execution of this policy by centralizing authority heretofore granted by law to several agencies and by granting additional authority with respect to interstate and foreign commerce in wire and radio communication.[21]

The FCC carries out these responsibilities for international telecommunications by approving the construction and operation of communications facilities, the offering of services, and the tariffs, or rates, charged therefore, by allocating and assigning radio frequencies to non-Federal Government users,[22] and by participating in international negotiations and conferences. The FCC also establishes rules and regulations for international telecommunications, has attempted to engage in facilities planning and coordination, and significantly influences U.S. policy and the reactions of foreign entities by virtue of the speeches and statements of the Chairman and Commissioners. The Communications Satellite Act of 1962[23] gives the FCC regulatory responsibilities over the common carrier activities of Comsat as well.[24]

The Commission's international functions have been divided among eight bureaus and offices: the Common Carrier Bureau, Office of Science and Technology, and the Mass Media Bureau, which perform the bulk of the international activities, and the Office of Plans and Policy, Office of General Counsel, Private Radio Bureau, Field Operations Bureau and Office of the Managing Director which are also involved. In 1981 an Assistant to the Chairman for International Affairs was appointed to coordinate FCC planning and activities, as well as to advise the Commissioners in international communications policy. An internal International Telecommunications Coordinating Committee also was created within the FCC "to assist in focusing the varying international functions of the involved Commission elements."[25]

United States Trade Representative

The United States Trade Representative is a Cabinet-level official with the rank of Ambassador who has responsibility for setting and administering overall international trade policy. The agency he heads (USTR) was first established in 1963 as the Office of the Special Representative for Trade Negotiations, and functions now as part of the Executive Office of the President.

Presidential Reorganization Plan No. 3 of 1979 (implemented by Executive Order 12188) has given USTR primary responsibility "for developing, and for coordinating the implementation of, United States international trade policy," saying that he shall serve "as the principal advisor to the President on international trade policy and shall advise the President on the impact of other policies of the United States Government on international trade," as well as having "lead responsibility for the conduct of international trade negotiations."[26] Whether the policy, advice, or negotiations are on trade in telecommunications equipment, telecommunications and information services, or any of the multitude of industries increasingly dependent on international telecommunications, the USTR is in the midst of international telecommunications policymaking.

While the Presidential Reorganization Plan establishing the Office of the United States Trade Representative uses terms such as "primary responsibility," "principal advisor," and "lead responsibility," it nonetheless assigns trade functions to the Secretary of Commerce, and states specifically as to the Secretary of State:

> Nothing in this reorganization plan is intended to derogate from the responsibility of the Secretary of State for advising the President on foreign policy matters, including the foreign policy aspects of international trade and trade-related matters.[27]

Elegant distinctions on occasion are drawn among the trade "facilitation," "promotion," and "negotiation" responsibilities of these players. Again, however, the question of overlapping responsibilities for international telecommunications policymaking is aggravated rather than resolved by this reorganization document. Nor has any serious attempt been made to reconcile the delegations of authority here with the potentially conflicting grants of authority found in other reorganization plans and Executive Orders, such as Executive Order 12046 discussed earlier.

Cabinet Council on Commerce and Trade

The Cabinet Council on Commerce and Trade is one of five Cabinet Councils formed by President Reagan in 1981. The purpose underlying the creation of the Cabinet Councils was to establish an orderly process for reviewing issues requiring a decision by the President, who acts as the Chairman of each Council. The Cabinet Councils were designed to operate as subgroups of the full Cabinet. The creation of the five Cabinet Councils has resulted in a rerouting of certain issues within the Federal bureaucracy. Issues formerly decided in OMB now receive consideration on a substantially higher political level with more White House control.

The Secretary of Commerce serves as the Chairman pro tempore of the Cabinet Council on Commerce and Trade. The other members include the Secretary of State, the Secretary of the Treasury, the Attorney General, the Secretaries of Agriculture, Labor, and Transportation, the U.S. Trade Representative, and the Chairman of the Council of Economic Advisers. The Vice President, the Counsellor to the President, and the Chief of Staff are ex officio members of the Council.

Although the Cabinet Council on Commerce and Trade was active the first six or eight months of its existence, with no definite schedule of meetings, it has been relatively inactive subsequently. The largest percentage of issues discussed in the Cabinet Council on Commerce and Trade concern trade; however, the focus has been on "smokestack" industries such as steel and automobiles rather than communications. The AT&T litigation and proposed domestic telecommunications legislation (S. 898 and H.R. 5158, 97th Congress) were issues discussed in the

Cabinet Council on Commerce and Trade, along with the role of the Postal Service in electronic mail, and proposals on home recording and cable copyright.

Also addressed but not resolved was the organization of the Executive Branch to deal with international telecommunications policy and the ambiguity of Executive Order 12046. This problem was passed to a working group of staff members to resolve, and no recommendation has been resubmitted to the Cabinet Council.

PROBLEMS

Executive Branch Policymaking

The traditional organization of policy development in the Federal Government encourages the labelling of issues as foreign policy, trade, telecommunications, spectrum, information, national security, etc. For example, foreign policy issues are handled by the Department of State; trade issues by the U.S. Trade Representative and by the International Trade Administration within the Department of Commerce; telecommunications by the FCC and NTIA; and so on.

International telecommunications issues, however, are typically complex and rarely so easily categorized. The Computer and Business Equipment Manufacturers Association (CBEMA) recognized this in its comments submitted in response to NTIA's November 2, 1982 Notice of Inquiry:

> The Executive Branch of the U.S. Government, as currently structured, is designed to deal with well bounded problems and policies in domestic communications, international trade in traditional raw materials and manufactured goods, and the legitimate "informational interests" of citizens and enterprises. However, it is ill-equipped to deal with problems and policies which cut across the boundaries of those areas or the agencies chartered to deal with them. The existing executive agencies lack the charter or experiential background necessary to deal with issues of the contemporary technology involving international channels for information, the information that flows in those channels, and the processes which might occur during transit.[28]

Choosing a label for an issue may well determine not only where in the Executive branch policy will be developed, but also what expertise and point of view will be applied in the process.

Former Under Secretary of State for Security Assistance, Science and Technology, Matthew Nimetz, summarized the difficulties inherent in the Executive Order 12046 when he, in 1980, explained to the Subcommittee on

Government Information and Individual Rights of the House Government Operations Committee the division of Executive Branch authority. After reviewing the overlapping delegations of authority in the Executive Order, Nimetz testified:

> So, I see two focuses. One is that we have primary responsibility for foreign affairs, and in negotiations. The Secretary of Commerce — under whose authority NTIA falls -- is the primary policymaker in the communications area. And we certainly defer to their authority and expertise. It is not the simplest area to describe. You have to go through the Executive Order rather carefully to parcel it out.[29]

The House Committee was less than satisfied. It responded in its Report:

> While State and Commerce are carefully parceling out their responsibilities, the private sector is understandably confused about where to go to ensure effective contributions to policy development and to get help for particular problems.[30]

Indeed the Subcommittee had received highly critical private industry testimony. For example, the Association of Data Processing Service Organizations (ADAPSO) wrote:

> The United States enters discussions, debates and participates in other fora involving international computer services with no real coordinated national policy and with many disjointed representations. As a result, independent American firms do not have the support of a unified national policy when they negotiate with foreign governments or PTTS. . . .[31]

Tymshare Incorporated told the Subcommittee of its lengthy difficulties dealing with the Japanese Government and the Japanese international telephone agency (the KDD) when trying to establish a computer services venture in Japan. They then described having "worked extensively with the numerous U.S. Government agencies apparently involved in the area of international information flow, in an attempt to get some useful assistance."[32] Their experience and resultant attitude is significant:

> Unfortunately, our disappointment and frustration due to the apparent lack of interest, assistance, and results from any U.S. Government agency was in some ways more frustrating than the agonies in dealing with the Japanese. We really had hoped and expected some form of support and assistance from our U.S. Government agencies. We were thoroughly disappointed by their reluctance to be involved and to take action but were most frustrated by the refusal of each agency to acknowledge that they had any responsibility or authority to provide such assistance.[33]

The fact that the forty-eight Federal Government delegates to WARC-79 were drawn from nine different agencies (FCC, NTIA and other parts of the Commerce Department, Defense, State, NASA, ICA, Transportation, the National

Science Foundation and the White House Office of Science and Technology Policy) further illustrates the broad range of concerns involved and the need for coordinating mechanisms in the U.S. Government to deal with international telecommunications issues. The Interagency Group on International Communications and Information Policy, described earlier, was created to try and deal with the lack of effective Executive branch coordination. Yet as the Report on S. 2469 by the Senate Committee on Commerce, Science, and Transportation recently noted, "there is no statutory or administrative basis for the interdepartmental group."[34] Its activities are not necessarily treated as priority matters by the participating agencies, nor given adequate high-level attention. Despite the international ferment in telecommunications issues in recent months, the Group has met only sporadically.

While issues before the Interagency Group may not receive enough high-level attention, the only other established mechanism for coordinating international policy development efforts, the Cabinet Council on Commerce and Trade, suffers from the opposite problem. Only the rare issue will warrant the study and attention of a multitude of Cabinet-level officers, and none will be able to have continuous monitoring, feedback, industry input, etc. over extended periods of time.

At the State Department, the former Under Secretary for Security Assistance, Science and Technology told a House hearing that he had "responsibility for ensuring a comprehensive U.S. international communications policy and ensuring close collaboration with other interested agencies."[35] At the same time, however, he announced a transfer of principal responsibility for transborder data flow problems affecting the interests of U.S. firms to the Bureau of Economic and Business Affairs, which is responsible to the Under Secretary for Economic Affairs. He also noted that "Other Bureaus will, of course, retain their lead roles in such matters as advanced technology, legal matters, and management of U.S. participation in international organizations."[36]

The House Committee on Government Operations responded that these Bureaus are scattered throughout the State Department, report to different authorities, and represent the U.S. in various forums. The attached organization chart of offices within the State Department concerned with international communications policy gives a picture of the complexity, overlapping responsibilities and diffused authority within that Department alone.

One of the areas where the lack of effective Executive Branch policymaking is raising the greatest concern is in preparation for international conferences and

Offices in the Department of State concerned with International Communications Policy *

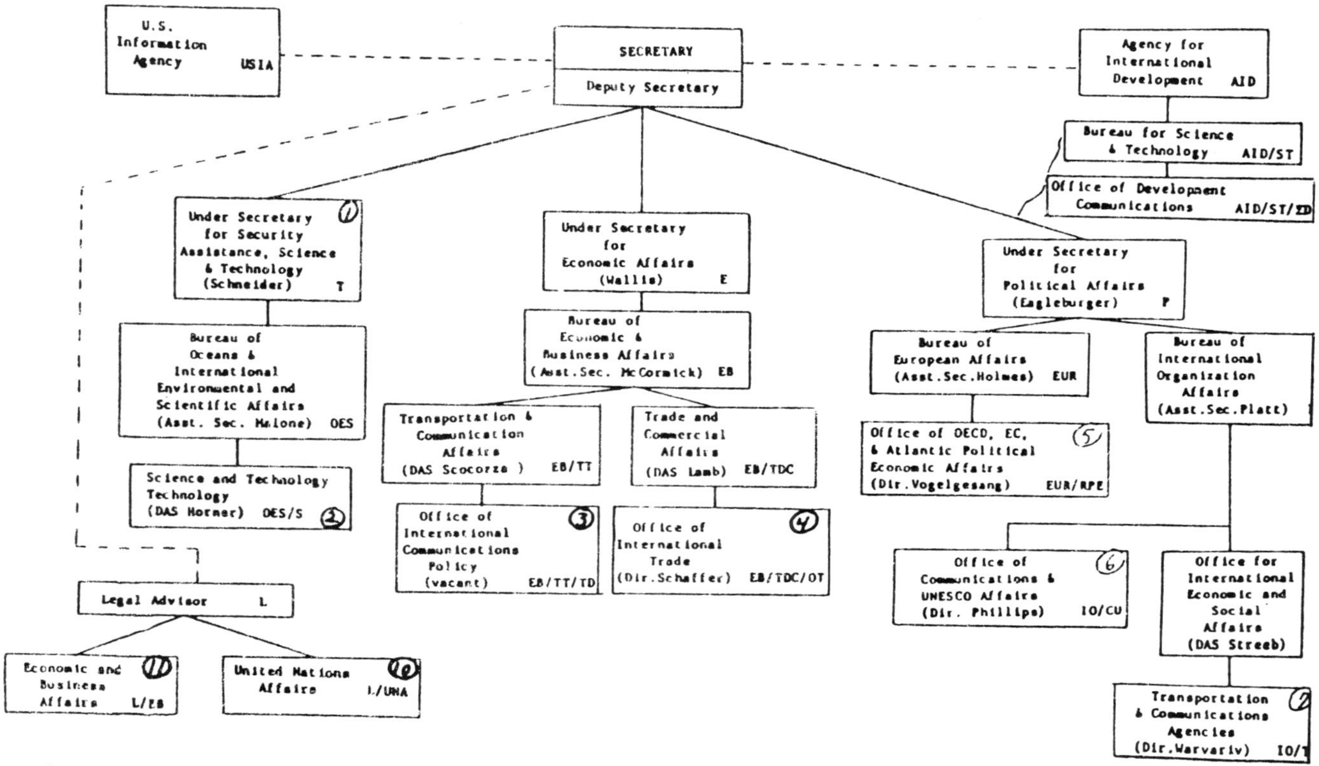

* Senate Comm. on Commerce, Science, and Transportation, International Telecommunications Act of 1982, S. Rep. No. 97-669, 97th Cong., 2d Sess. 18 (1982).

negotiations. Vital U.S. interests are at stake in meetings of organizations such as the ITU, UNESCO, and the OECD. Yet as another section of this report details, the U.S. Government seems ill-prepared to meet the challenges presented.

In view of the international telecommunications policymaking structure of the Executive Branch, both between and within agencies, it is not surprising that private industry is often confused. Spokespersons from the private sector cite the fragmented, ambiguous grants of authority, and consequent problems of lack of coordination, myopia, and lack of accountability. They call for an ultimate arbiter of U.S. policy, with sufficient resources of both expertise and power, who can then serve as a focal point for private sector input and for foreign negotiations. In recognition of their vital importance to the nation, international telecommunications policy decisions require high level attention in the Executive branch.

The Executive Branch and the Independent FCC

Structural deficiencies within the Executive branch have long been discussed, and many remedies have been proposed to improve the development and implementation of U.S. international telecommunications policy. Another structural "flaw," with us since the Communications Act of 1934, however, has only recently become significant, or at least been recognized as such. Irrespective of any policy developed within the Executive Branch, the FCC, independent of direct Presidential control, may effectively establish international policy on its own, and may advance or thwart Administration policies. The FCC was established by the Communications Act of 1934, in part, ". . . for the purpose of securing a more effective execution of this policy by centralizing authority heretofore granted by law to several agencies. . . ."[37]

But while the need for centralized authority was recognized even then, the Congress, nonetheless, carefully divided responsibility for allocating the radio frequency spectrum between the FCC for non-Federal Government users and the President (since delegated to NTIA) for Federal Government users. See 47 U.S.C. Sec. 305.

As a practical matter, responsibility for international telecommunications policymaking also has been divided between the FCC and the Executive branch, although with no clear dividing lines such as those which exist for frequency allocation authority. The FCC has increasingly taken an affirmative policymaking role, not one limited to its regulatory or adjudicatory model, in part because of the failure of the Executive branch itself to develop and implement policy in a coherent manner. Control Data Corporation, for example, has written in terms of the FCC acting "to fill the vacuum" left by the failure of other Government entities, even though some such actions "are generally considered to be outside the statutory scheme set forth by the Communications Act of 1934, and cannot lead to the resolution of existing problems in the international telecommunications and information flow area."[38]

As international telecommunications has grown dramatically, so too has the significance of FCC policy decisions, not merely for telecommunications, but also in their impact on national security, trade, and foreign policy. Therefore, as the General Accounting Office reported in its recent analysis of the FCC's international telecommunications activities:

> Under its normal process for authorizing communications facilities, FCC considers factors such as technological development and consumer economies that can result from these developments. However, in several proceedings over the last few years, FCC has had to go beyond its traditional areas of expertise to consider foreign affairs, national security, and U.S. trade policy.[39]

The GAO cited as just one example the facilities authorization proceeding in which the FCC reviewed AT&T's proposed award of a contract for fiber optic cable to Western Electric instead of to a lower-bidding Japanese firm. The Departments of Defense, State, and Commerce, and the USTR, among others, discussed the national security, foreign policy, and international trade issues at stake, but it was the FCC which had final authority to decide the matter.

In carrying out his foreign policy functions with respect to telecommunications, the Secretary of State is required by Executive Order 12046 to "coordinate with other agencies as appropriate," and in particular to "give full consideration to the Federal Communications Commission's regulatory and policy responsibility in this area."[40] No similar statutory or other requirement mandates that the FCC must take foreign policy, trade, and national security concerns into account, nor even that it must seek or consider the views of the Executive Branch agencies whose primary functions these are Concerns such as national security, of course, are generally considered important components of the broad "public interest" mandate under which the FCC operates. The statute gives the FCC little clear guidance, however, concerning the decisional weight it should accord such concerns. Rather, the FCC is obliged on an ad hoc basis to endeavor to balance Executive branch needs (as to which it lacks full knowledge) with a diversity of other matters, a process often as frustrating to the FCC as to the pertinent Executive agency.[41]

OPTIONS

Maintain the Status Quo

Despite the preceding review of some of the problems and criticisms of the organization of the U. S. Government to deal with international telecommunications policymaking, one obvious option which must be considered by Congress and the President is maintaining the present structure.

Telecommunications policymaking responsibilities are fractionalized, with consequent problems of insufficient planning, coordination, and final decision-making authority. The present system is clearly inadequate to meet current needs. However, given the diversity of institutions and Congressional committees involved, not to mention private sector constituencies, the transaction costs of major changes obviously could prove substantial.

Executive Branch Policymaking

Increased Coordination. The focus of the analysis thus far has been on the structure of the Government, looking at the multiple and overlapping grants of authority both to various Government agencies and within those agencies. Many have expressed concern not only with the number and diversity of agencies interested in international telecommunications and information issues, however, but rather with the general lack of adequate coordination, communication, and cooperation among those agencies. The formal mandates of the Secretaries of Commerce and State and of the USTR, among others, all require coordination of efforts, but without setting forth specific mechanisms by which this is to occur.[42] As a practical matter, all too often coordination falls victim to questions of jurisdictional disputes between departments and agencies, lack of adequate high-level attention, varying agency priorities, time pressures, and lack of resources.

Expressing what has become a commonly-held point-of-view, the Computer and Business Equipment Manufacturers Association told NTIA:

> CBEMA believes that until such time as there is a central authority in the United States vested with the responsibility of coordinating and articulating overall international communications and information policy, negotiating international rules with our trading partners, and participating in the development of appropriate means by which those rules will be enforced, the U.S. will continue to be of a disadvantage.[43]

Alternative proposals which could be considered to overcome the lack of Executive Branch coordination and cooperation include:

(1) Formalizing the present Interagency Group on International Communications and Information Policy and strengthening its mandate.

(2) Setting up another interagency council, task force, or body headed by either the Secretary of State, the Secretary of Commerce, or the USTR, or their senior level designee.

(3) Charging the President's National Security Adviser, OMB, or another Executive Office entity with coordinating interagency efforts.

(4) Designating a Special Assistant to the President for International Communications Policy with a small professional staff.

(5) Giving all responsibility for international communications policy to either the Department of State or Department of Commerce.

(6) Creation of a Department of Communications.

Each alternative is designed to establish the specific process and to hold a specific individual responsible for the coordination which broad agency mandates have not accomplished. Each would provide a focus for private sector input. Yet each seeks consensus, which may translate to the lowest common denominator and produce inherently weak policies, or fails to address what occurs in the absence of consensus. Will the agencies continue to speak with more than one voice, so that the Congress, the FCC, U.S. industry, and foreign entities are unsure what Executive Branch policy is? There must be established within the Federal government by whatever mechanism or structure a focal point for the handling of international communication matters. At the current time, the private sector, foreign governments, and even the U.S. government itself are not altogether clear as to which part of the Federal government deals with these matters and has the authority to express the views of the United States.

Consolidation of Authority. In reviewing the record to date, one could argue that efforts merely to improve coordination and communication among the many diverse Government entities with a hand in international telecommunications policymaking are superficial and doomed to failure because they fail to deal with the underlying cause of the problem. What is missing is not coordination, but coordinated decisionmaking, the development of a single Executive Branch position on an issue which is then accepted and followed throughout the Administration. This requires not only a person charged with obtaining various agency views and drawing on agency expertise, but also with the specific authority to arbitrate differences and finally determine the Administration's policy, a policy which individual agencies are not free to contradict, ignore, or undermine.

The guiding principles in establishing such authority are:

(1) the authority must be centralized in a single place;

(2) the coordinating entity, agency or person must have a clear and strong grant of authority and responsibility for coordinating international telecommunications and information policy;

(3) the entity, agency or person must have the power to mediate differences among agencies and make final decisions;

(4) the entity should establish a formalized interagency policy advisory body, supported by a secretariat, and which meets regularly to assist in the formulation and implementation of policy.

(5) there should be a clear and regular structure for industry input.

In addition, the coordinating entity should be responsible for representing the U. S. at all international conferences dealing with telecommunications and information issues, and the head of the entity should be clearly identified as the President's principal adviser on international telecommunications and information issues.

Private sector input is also a critical element in the development of effective and responsive policy. In the view of Michael R. Gardner, Chairman U.S. Delegation, Nairobi Plenipotentiary Conference:

> A Council on International Telecommunication should be established by the White House consisting of ten to twelve leaders from diverse segments of the telecommunication industry. This Council should have as its primary goal the task of forging a new and much needed alliance between the U.S. telecommunication industry and segments of the federal government appropriately involved in international telecommunication. By forging a more open and meaningful partnership between industry and government, a new form of entrepreneurial diplomacy should guide the joint activities of government and industry in the international policy fora and the worldwide marketplace for telecommunication.

While many agree with the need to consolidate final decisionmaking authority, there is no consensus as to where in the Executive branch that authority should be vested. Within existing institutional structures, potential designees include the Secretary of State, the Secretary of Commerce, the President's National Security Adviser, and the U. S. Trade Representative. Consideration can also be given to establishing a new Special Assistant or Adviser to the President, or a new Board or Office within the Executive Office of the President, or even a Department of Communications. Each would bring a different expertise and perspective to the issues.

Recognizing the previously discussed problems of labeling issues as merely foreign policy or trade or national security or telecommunications, however, the question must be raised as to whether even the granting of final decisionmaking authority is sufficient. Will the designated official have sufficient access to the diverse expertise housed in other agencies in order to exercise that decisionmaking responsibility wisely? Or must adequate staff expertise on the array of interrelated issues be consolidated along with the final authority?

Having been granted responsibility of preparing for and representing the United States at international negotiations, this centralized authority will require the additional staff to meet the crowded international conference agenda which has already been set. A strong case has been made for the establishment in Government of a permanent conference preparation staff, whether consolidated at one agency or drawn on a continuing basis from existing entities. The skills, knowledge of the issues and players, and the continuity of representation, all under the control of a high level political appointee, would greatly strengthen U.S. capability to achieve the goals set for international negotiations.[44]

A possible way to implement the approach described above is indicated in the following diagram. By clearly designating central accountability and by coordinating interrelated policy objectives, strategies, and resources on a systematic and regular basis, U.S. actions internationally will be mutually supportive and better achieve U.S. goals.

The senior level policy authority would be supported by a policy advisory body comprised of concerned and affected Government agencies, thus providing a full opportunity for policy input from all interested agencies.

From the senior level policy authority, the advisory body would receive broad policy guidance on the Administration's goals and objectives for international telecommunications. The policy authority might also task the advisory body, where policy voids exist, to make appropriate policy recommendations. In this way, the advisory body would be responsive in a timely manner to any changes in the Administration's positions and such information would be disseminated quickly and comprehensively to the appropriate activities. The advisory body, in turn, would refer for resolution and guidance to the senior level authority those policy issues which appear to elicit differing or conflicting positions on the part of advisory body members.

The advisory body would meet on a regular basis, possibly twice a month. It would have a small professional staff to oversee policy development activities and a secretariat to support its distribution of papers and agenda items among the

EXAMPLE

EXECUTIVE BRANCH STRUCTURE COORDINATED DECISIONMAKING

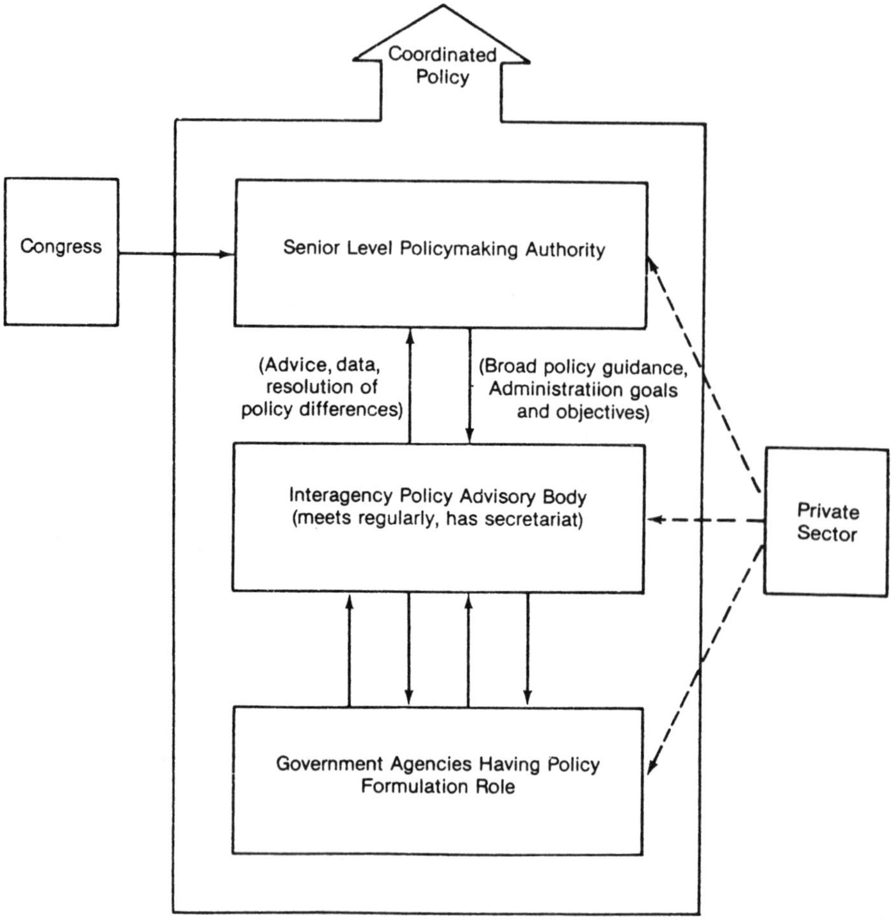

membership. The Secretariat would also keep consolidated and centralized records of advisory body actions and policy statements. Its terms of reference should be broad, to encompass at a minimum, all of the major interrelated telecommunications issues identified by NTIA in its Notice of Inquiry.

Some channel would also be established to take note of the views of the Legislative branch and private sector and facilitate meaningful interaction in the policy formulation process. The private sector could have access in a number of ways. The senior level policy authority might establish an advisory group or the private sector could organize itself into a telecommunications council and make its views known on issues of importance or concern. Contact with the Legislative branch could be in the form of periodic briefings and discussions.

The Executive Branch and the Independent FCC

The options discussed above may solve or alleviate structural deficiencies within the Executive branch. If they yield an Executive branch speaking with one voice, they will also have solved a major problem for the FCC, which is often confronted with competing, if not conflicting, views of individual federal agencies. Today the FCC, without the mandate or expertise to do so, must balance concerns of national security versus trade policy, or foreign relations versus balance of payments effects. Instead, the FCC should be presented with a single Administration position which already represents the trade-offs and cost/benefit analyses which are the Executive's function to make.

Given a single, clear representation of the Administration's views on the foreign policy, security, trade, or other implications of contemplated action, however, the FCC is still under no legal mandate to consider the Administration's position, let alone defer to it. Nor is the FCC specifically required to consider the foreign policy or trade implications of its decisions. Legislation is possible to require the FCC specifically to weigh these factors in its deliberations and to require consideration of the views of the Executive branch agencies. It would be very difficult, of course, to measure the impact of such requirements on the outcome of future FCC deliberations. To provide greater assurance of Executive branch influence or control might require the establishment of a Presidential veto power over FCC international telecommunications actions.

The first question in exploring a grant of authority to the President to override FCC decisions is what are the appropriate reasons for the exercise of the veto. Possible grounds for intervention include national security, foreign policy, international trade, economic well-being, or any combination thereof. Second, within what time limits must the President act? Clearly the President cannot be expected to review each proceeding nor would the marketplace be able to function if every FCC order were left as only tentative. Setting an automatic effective date of 15 or 30 days after the FCC issues an order -- unless the President takes affirmative action -- would provide sufficient certainty. Finally, a decision must be made on the extent to which the power to exercise the veto may be delegated. Delegation could again spark battles over "turf" and control in the Executive branch, and if multiple grounds for intervention are provided, the risk would exist of having many Executive agencies all looking over the FCC's shoulder. It is also important to note that a veto power can only void an FCC decision. It cannot modify the decision nor deal with a failure to act.[45] To do so would require affirmative policymaking authority in the Executive branch.

Another option, therefore, is to transfer the bulk of the FCC's international telecommunications policymaking responsibility into the Executive branch. The execution of that policy, such as the issuance of licenses or the approval of facilities within guidelines set by the Administration, could remain at the FCC. The underlying policy formulation, and the consideration of issues such as national security, foreign policy, trade, and economics, however, would be within the control of the President. Without a controlling Executive branch statement of policy, the FCC would be powerless to act. New policy initiatives thus could not be undertaken by an independent agency which was not subject to direct Presidential oversight. Many of the traditional regulatory criteria applied by the FCC have now been dwarfed in significance in the international area by concerns such as security and trade. Therefore, the traditional public utility regulatory model may no longer be the appropriate one for international telecommunications policymaking, and the FCC may no longer be the proper situs for that responsibility.

The various organizational and operational options discussed in this chapter are graphically represented in the following charts.

THE EXECUTIVE BRANCH AND THE INDEPENDENT FCC

STATUS QUO

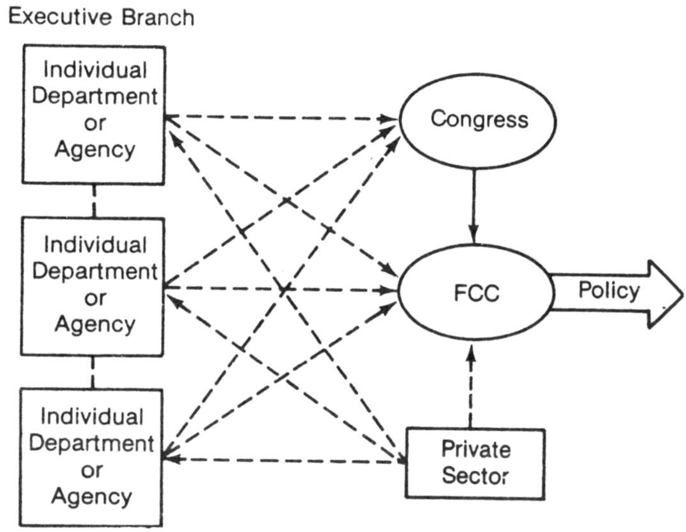

COORDINATED EXECUTIVE BRANCH DECISIONMAKING

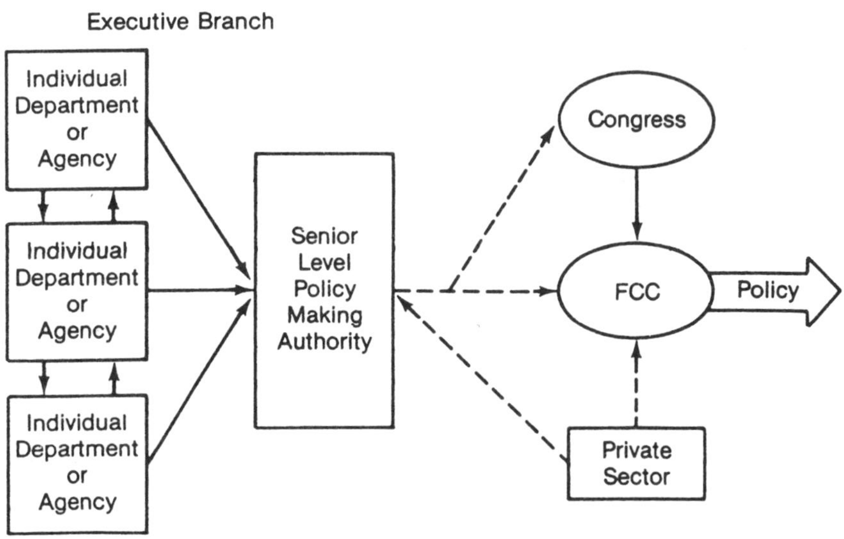

PRESIDENTIAL VETO

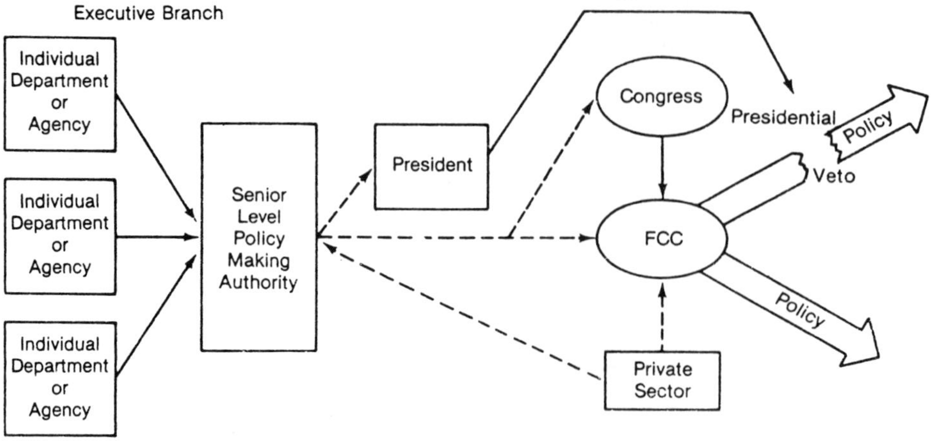

EXECUTIVE BRANCH POLICYMAKING, FCC IMPLEMENTATION

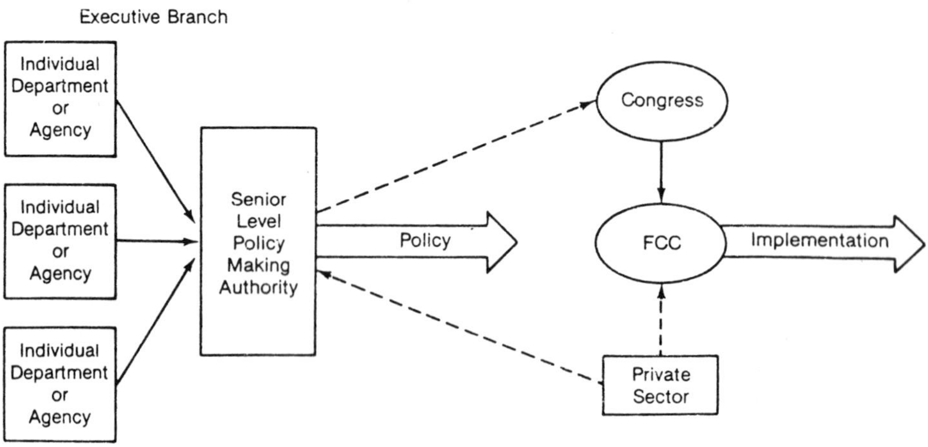

NOTES TO CHAPTER FOUR

[1] A nonexhaustive list includes: Department of Commerce, Department of State, United States Trade Representative, Federal Communications Commission, Department of Defense, National Security Council, Office of Science and Technology Policy, Office of Management and Budget, Department of Justice, United States Information Agency, Board for International Broadcasting, National Aeronautics and Space Administration, Department of Transportation, Department of the Treasury, Department of Energy, United States Postal Service, Postal Rate Commission, Federal Reserve System, National Science Foundation, General Services Administration, Small Business Administration, Office of Technology Assessment, General Accounting Office, International Trade Commission, International Development Cooperation Agency.

[2] Senate Comm. on Interstate Commerce, *Investigation of International Communications by Wire and Radio, Interim Report*, S. Rep. No. 1907, 79th Cong., 2d Sess. 3 (1946).

[3] *Id.*

[4] *Id.* at 4.

[5] President's Communications Policy Board, *Telecommunications, A Program for Progress*, at 9 (1951).

[6] *Id.* at 184.

[7] *Id.* at 10.

[8] *Id.* at 11.

[9] President's Task Force on Communications Policy, *Final Report*, ch. 9, at 2,10 (1968).

[10] Allocation and assignment of spectrum space to all non-Federal entities is the responsibility of the FCC.

[11] *See* Executive Order 12046, 3 CFR 158 (1978 comp.).

[12] 47 U.S.C. § 721(a).

[13] 3 CFR 197 (1977 comp.).

[14] 3 CFR 158 (1978 comp.).

[15] See discussion *infra.*

[16] Executive Order 12046, §§ 5-201, 5-202, 3 CFR 158, 164 (1978 comp.).

[17] *Id.* § 2-404 at 161.

[18] 1 Foreign Affairs Manual 145 (h)(3) (Dec. 29, 1981).

[19] 47 U.S.C. §151 *et seq.*

[20]See Omnibus Budget Reconciliation Act of 1982, Pub. L. No. 97-253, § 501 (1982).

[21]47 U.S.C. § 151.

[22]The allocation and assignment of spectrum space to Federal government entities is performed by NTIA under a delegation of Presidential authority.

[23]47 U.S.C. §701 et seq.

[24]See discussion infra.

[25]General Accounting Office, The Federal Communications Commission's International Telecommunications Activities, Report to the Chairman, House Subcomm. on Government Information and Individual Rights at 6 (April 19, 1982).

[26]3 CFR 513 (1979 comp.).

[27]Id. at 516.

[28]Computer and Business Equipment Manufacturers Association, Comments in Response to National Telecommunications and Information Administration's Notice of Inquiry of Nov. 2, 1982, at 4 (Dec. 3, 1982).

[29]House Comm. on Government Operations, International Information Flow: Forging a New Framework, H.R. Rep. No. 96-1535, 96th Cong., 2d Sess. 43 (1980).

[30]Id. at 44.

[31]Letter from Jerome L. Dreyer, Executive Vice President, ADAPSO, to Hon. Richardson Preyer, Chairman, House Subcomm. on Government Information and Individual Rights, May 30, 1980, in International Data Flow, Hearings before the House Subcomm. on Government Information and Individual Rights, 96th Cong., 2d Sess. at 703 (1980).

[32]Statement of Warren E. Burton, Vice President, Tymshare, Inc. before the House Subcomm. on Government Information and Individual Rights, March 10, 1980, in International Data Flow, Hearings before the House Subcomm. on Government Information and Individual Rights, 96th Cong., 2d Sess. at 75 (1980).

[33]Id.

[34]Senate Comm. on Commerce, Science, and Transportation, International Telecommunications Act of 1982, S. Rep. No. 97-669, 97th Cong., 2d Sess. 17 (1982).

[35]Testimony of Matthew Nimetz, Under Secretary of State for Security Assistance, Science and Technology before the House Subcomm. on Government Information and Individual Rights, March 27, 1980, at 13.

[36]Id.

[37]47 U.S.C. § 151.

[38]Statement of Philip C. Onstad, Control Data Corporation, before the House

Subcomm. on Government Information and Individual Rights, March 10, 1980, in International Data Flow, Hearings before the House Subcomm. on Government Information and Individual Rights, 96th Cong., 2d Sess. at 53 (1980).

[39] General Accounting Office, The Federal Communications Commission's International Telecommunications Activities, Report to the Chairman, House Subcomm. on Government Information and Individual Rights at 3 (April 19, 1982).

[40] Executive Order 12046, § 5-201, 3 CFR 158, 164 (1978 comp.).

[41] As the State Department told the Congress in this regard:

> As an independent regulatory agency, the FCC is not in a position to determine the foreign policy and national security implications of plans for international facilities and services.[47]
>
> Letter from J. Brian Atwood, Assistant Secretary of State, to Hon. Richardson Preyer, Chairman, House Subcomm. on Government Information and Individual Rights, August 21, 1980, in International Data Flow, Hearings before the House Subcomm. on Government Information and Individual Rights, 96th Cong., 2d Sess. at app. 4 (1980).

[42] See, e.g., Sections 2-404, 2-405 and 5-201 of E. O. 12046 and Sections 1(b)(1) and 1(b)(3) of Reorganization Plan No. 3 of 1979.

[43] Computer and Business Equipment Manufacturers Association, Comments in Response to National Telecommunications and Information Administration's Notice of Inquiry of Nov. 2, 1982, at 5 (Dec. 3, 1982).

[44] See the chapter on U.S. Participation in the ITU and preparation for radio conferences, infra.

[45] For a detailed discussion of the use of a Presidential veto in the facilities planning context, see the chapter on Facilities and Networks, infra.

Chapter Five

FACILITIES AND NETWORKS

International telecommunications facilities -- chiefly undersea cables and geostationary satellites — link the United States with overseas networks and are owned jointly by United States carriers and foreign telecommunications administrations. Each generally owns half of each link extending to the imaginary midpoint for undersea cables and from earth station to spacecraft in the case of satellite circuits. Establishing international telecommunications facilities is thus a cooperative undertaking involving entities of two or more nations -- private, regulated firms in the case of the United States, and the government or government-designated monopolies in other countries.

Satellite planning for most international communications (capacity, cost, quantity, etc.) is undertaken through the International Telecommunications Satellite Organization (Intelsat), which is jointly owned and administered by operating entities of 108 nations. Undersea cable systems for use by the United States are planned for by the American Telephone and Telegraph Company in consultation with United States international record carriers (e.g., Western Union International, RCA Global Communications, Inc., ITT World Communications, Inc., TRT Telecommunications, Inc., FTC Communications, Inc.) and the relevant foreign telecommunications authorities.

This chapter reviews the major issues involved in the establishment of international facilities and networks by focusing on:

- Allocation of spectrum resources;
- Allocation of satellite orbital resources;
- Facilities planning and authorization;
- Comsat and Intelsat issues; and
- Integrated Services Digital Networks (ISDN).

BACKGROUND

The last two decades have witnessed rapid development of

telecommunications and information technology. Increases in capacities and lowered unit costs are the most obvious results of these technological advances.

The first modern transoceanic submarine cable capable of providing voice services was placed in operation across the North Atlantic in 1956 (TAT-1). It was capable of providing only 36 basic 4 kHz two-way voice circuits. The latest North Atlantic submarine cable, now under construction, is TAT-7 with a capacity of over 4,000 voice circuits. It is expected that the next generation of submarine cables, using fiber optic and digital technology, will be available for transoceanic use in the late 1980s. This digital cable should be capable of providing (in a basic configuration) over 12,000 voice grade circuits, or a mixture of voice and television transmission. Conversation "throughput" is expected to be increased several-fold through the use of more efficient speech encoding and interpolation techniques.[1]

The technology of international satellites has also developed rapidly. The first commercial international communications satellite, known as Early Bird, was launched by NASA for Intelsat in 1965 and had a total capacity of 240 voice grade circuits or one TV channel. Each of the current, or fifth generation Intelsat satellites can be configured to provide more than 12,000 voice grade circuits plus two TV channels. The capacity and flexibility of these satellites will be enhanced by advanced earth segment technology such as time division multiple access (TDMA), compandors, and speech interpolation systems.[2]

Flexible regulation of the radio frequency spectrum and of the geosynchronous orbit (GSO) is necessary on an international scale to ensure the viability of both domestic and international satellite systems. The growth in technology has fostered continuing increases in efficiency, both in the use of the spectrum and the GSO. It is vital to U.S. national interests to assure international regulatory features are adopted that support efficient use of the GSO, guarantee equitable access, and ensure the orderly introduction of new technologies to the benefit of all users.

A recurring problem is the obsolescence of existing U.S. legislation in light of (1) the increasing importance of international telecommunications and information to the nation's economic well-being and national interest, and (2) the changes that permit transformation from regulated monopoly conditions to full and fair competition in a deregulated environment.

MAJOR ISSUES IN THE ESTABLISHMENT OF INTERNATIONAL FACILITIES AND NETWORKS

Allocation of Spectrum Resources

Allocation of the electromagnetic spectrum for specific radio services is accomplished at periodic World Administrative Radio Conferences (WARC) and Regional Administrative Radio Conferences (RARC) of the International Telecommunication Union (ITU). The ITU allocation table divides the world into three regions and makes a basic distribution of the frequency spectrum up to 275 GHz.[3] This spectrum is divided into 544 separate frequency band allocations. These allocations are made to 37 different radio services on either an exclusive or shared basis.

The basic allocation table was reviewed in 1959 and again during the 1979 WARC. The ITU allocation table may be changed in limited ways as a result of specialized conferences held more frequently, when the terms of reference permit such action. Most specialized conferences, however, are devoted to development of specific details for frequency use by a specific radio service and work within the framework of the overall allocation table.

The U.S. National Table of Frequency Allocations is a separate allocation table for use in the United States. This document follows the general framework of the ITU Allocation Table to the extent unique U.S. operations permit. In those instances where the U.S. allocation table differs from the international table, operations must be on a non-interference basis relative to other administrations operating in accordance with the ITU allocation table or in accordance with bi- or multilateral operating agreements. As a matter of practice, the United States deviates from the ITU allocation table only where operations are unlikely to cause interference outside of U.S. borders.

Economic Significance of Spectrum Allocation Tables. The telecommunications equipment manufacturing industry obviously has a vital interest in the allocation of radio spectrum on a national and international basis as it determines both potential demand for products and the cost of the equipment involved. When new radio equipment approaches the developmental stage, the new system must be designed to operate in a frequency band allocated to that particular radio service. If the system is to be marketed worldwide, it must be designed to operate in internationally allocated frequency bands. This presents major economic considerations because the wider the area of the world where a new product can be used, the greater the potential market, the larger the production quantities, and, in many cases, the lower the cost of each unit produced. This prospect of lower costs influences buyers and sellers and produces an incentive to strive for spectrum allocations which are consistent worldwide. Due to these factors many ITU disputes over spectrum allocation issues relate to how much spectrum in what frequency range a particular radio service will be allocated, as well as uniform applicability to all regions.

Political Aspects of Spectrum Allocation Tables. Different nations clearly have different internal and international telecommunications needs; they tend to be guided by their internal needs when negotiating for spectrum allocations at international conferences. Additional problems arise when the spectrum requirements for national security operations are the basis for a particular allocation.[4]

Currently, the use of spectrum on an international basis is accomplished through specific coordination and notification procedures. Frequencies are registered to whomever first coordinates the use with other countries for a particular frequency at a designated location and subsequently notifies the International Frequency Registration Board (IFRB). This process has satisfied the needs of most administrations, although questions have arisen concerning the complexity of the process by those with limited spectrum management resources. Questions have also been raised by some lesser developed countries concerning the ability of the present process to accommodate their future needs which, in turn, has stimulated interest on their part in long-term "a priori planning" for the spectrum and GSO.

Current Policy on Allocation of Spectrum Resources. U.S. policy for international spectrum management takes into account a large number of specific policy objectives. These objectives fall into several categories.

(1) The first category of objectives involves obtaining international recognition and protection of new and existing radio systems, as for example, the new U.S.-developed Global Positionary System (GPS), which has the potential of replacing a number of existing radio navigational aids.

(2) The second category involves the adoption of frequency allocations that realistically advance our economic and national security goals.

(3) The third category involves the introduction of spectrum conservation measures to permit more efficient use of the radio

spectrum, as for example, the introduction of single side band modulation in the aeronautical and maritime services.

(4) A fourth category of U.S. policy objectives concerns the conduct of future spectrum planning conferences, which will further refine how some radio services will be operated in the future.

The last category has received significant attention on the part of Congress and the public given that a number of developing countries are now seeking detailed frequency planning which would reserve spectrum assets for their future use. This approach is wasteful of spectrum assets and would inhibit technological progress.

U.S. preparation for international conferences — the process of developing concrete U.S. spectrum allocation proposals -- involves developing comprehensive information on the current use, needs, and rate of development of radio services, assessing the state of the art, and reviewing the operational practices used by these services. The ability of radio services to share allocations must be under constant evaluation if the most efficient applications are to be fostered. When U.S. requirements for international telecommunications systems or allocations are presented, coordinating these proposals with other nations materially enhances the likelihood of their acceptance.

Since the International Spectrum Allocation Table is reviewed in its entirety only about every two decades, the task of forecasting detailed spectrum requirements is a substantial undertaking. As the rate of change in the state of the art continues, this challenge becomes more acute. The consequences of faulty projections can have a serious impact on the future of many different radio services.

Recommendations on Allocation of Spectrum Resources. U.S. policy seeks to ensure timely application of the latest technology to minimize transmission bottlenecks and spectrum scarcity. The policy of allocating the spectrum in response to demonstrated economic, national security, and foreign policy needs and future requirements is most effective and should be maintained. Management of these allocations should be grounded on sound procedures that afford users flexibility, equitable access to meet operational needs, and facilitate the systematic introduction of new technologies beneficial to all.

The overall importance of planning for future radio spectrum requirements, both nationally and internationally, cannot be overstated. Service allocations and management procedures are the keystone to the entire spectrum management structure. The United States thus must continue to improve its overall ability adequately to prepare sound international proposals and to obtain international support. In few other areas of spectrum management is it more difficult to correct mistakes. A full discussion of methods to improve U.S. performance in attaining its goals at future scheduled radio and other ITU conferences (such as DBS-RARC-83, Mobile WARC-83, HF Broadcasting WARC-84, Space-WARC 85/88) is presented elsewhere in this report.

Allocation of Geostationary Orbit Resources

During the last five years there has been much discussion concerning "equitable access" to the geostationary orbit (GSO). The twentieth anniversary of the orbiting of the first successful geostationary satellite, NASA's Syncom II, will be marked in 1983. In these two decades there has been and continues to be a remarkable growth in the number and diversity of types of satellites which have been placed in orbit.

Most satellites relaying communications are located in the geostationary orbit. These satellites have created an institutional and structural revolution in the field of telecommunications and information, both domestically and internationally. The global connectivity made possible by satellites has had a fundamental impact on the practical delivery of communications.

Although many space services utilize the geostationary orbit, the Fixed Satellite Service (FSS)[5] is the largest in terms of total satellites. Dramatic growth in GSO use has occurred over the past 15 years, particularly in telephone, data transmission, and video distribution. This growth has occurred largely in developed countries, although developing countries have shared in this growth through their participation in Intelsat, through special leasing arrangements, or through construction of their own satellites. In 1981, Intelsat served 160 countries, territories, or possessions either directly or indirectly. Several developing countries (e.g., Mexico, Brazil, Indonesia, India, and China) are already operating or have firmly planned domestic satellite systems. Regional satellite systems, such as Arabsat, have planned launch dates, and interest in an African regional satellite system has been expressed. In addition, Intelsat is considering plans to offer non-preemptable domestic leased satellite services. This activity exemplifies the extensive interest and growing demand for use of the GSO by developing countries.

The current method of orbital assignment in the ITU for both geostationary and non-geostationary satellites is similar to but more flexible than procedures used for international notification of radio stations on earth. These procedures are used to obtain international interference protection of the satellite, to avoid interference to other satellite networks, and to inform administrations of others' plans. The procedures are technical in nature and involve parameters including satellite location, transmitter power, operating frequencies, antenna coverage, and receiver sensitivity.

The satellite notification process involves three phases — advance publication, coordination, and notification. In the first phase, the ITU's IFRB circulates to all administrations the information submitted by a country on its planned satellite network. After a six month comment period, the proposing administration coordinates with others to resolve any potential interference problems. After an additional six months and the resolution of all potential interference problems, the proposing administration submits notification of the satellite operation to the IFRB. This notification is an administration's "license" or "assignment" to use the orbit (and frequencies), and forms a basis for commenting on potential interference by the planned networks of other administrations.

A first step in considering potential orbit allocation scenarios involves an assessment of available GSO capacity. In order to provide a common baseline and because analog TV and telephony are the two most common uses of the FSS, system capacity is most commonly measured in equivalent 40 MHz transponders, which relate to one TV or 1000 equivalent voice grade channels. This measure of orbital capacity represents substantial simplification. It provides a reasonable means of estimating the potential for orbital crowding, however, since it understates actual achievable capacity for modern systems.

In analyzing the subject of the total capacity of the geostationary orbit, it is instructive to estimate current in-orbit capacity along with an estimate of its present usage. As a means of providing some insight into this information, two elements or blocks of orbit capacity now being used are:

(a) the Intelsat system, and

(b) the systems for the Western hemispheric arc to provide fixed domestic satellite service to the U.S., Canada, and Latin America, which have been, or are in the planning process.

Japan, Intelsat, the United States, and the Soviet Union have undertaken analyses of the capacity of the GSO. The results of some of these analyses show:

- Achievable capacity is very great, something in the order of 5 to 20 x 10^6 telephone channels, and this could be increased if standards were adopted which maximize satellite orbit capacity in relation to service area.

- The U.S.S.R. study estimates the maximum theoretical capacity in television channels to be in the order of 400 to 1,176 per degree of orbit.

These estimates have all been based on certain technical assumptions, all of which are practical and many of which are in operating systems. The analyses mentioned above have indicated that there are certain directions in which satellite system design and operation should evolve if further orbit capacity is to be achieved. These include the following:

- o The use of cross polarization to achieve frequency re-use;
- o Increasing the number of areas served from a given orbit location;
- o Increasing use of limited coverage beams;
- o Better control of satellite and earth station antenna side-lobes;
- o Further improvement in satellite station keeping; and
- o Increase in intersystem noise allowance.

Current Situation concerning GSO Allocation. There have been comments and observations regarding possible "crowding" of the GSO. These comments have fueled fears, particularly among developing countries, that no orbital slot will be available at such time as they may launch their own satellites. At the 1979 WARC, the LDCs proposed and obtained concurrence on a resolution to convene a conference to "guarantee in practice for all countries equitable access to the geostationary satellite orbit and the frequency bands allocated to the space services."[6] The 1982 ITU Plenipotentiary scheduled sessions for 1985 and 1988 on this topic. Many technologists and policymakers in the developed nations believe:

- o There are a number of techniques and improvements which can be implemented over time to increase availability of capacity dramatically.
- o No satellite system in the relevant frequency bands has yet been denied access to the GSO.
- o The United States, in its domestic regulatory procedures, is breaking ground in forcing the implementation of improved orbit utilization technology techniques.
- o The actual number of satellites which may be accommodated in the GSO in any particular frequency band is very large and is being increased constantly through the use of improved technology.
- o Effective management procedures that have the flexibility to take into account changing requirements, technology, and operational arrangements offer the best means of ensuring both efficient use of and equitable access to the GSO.

The fact that several countries located on the equator have claimed sovereignty over the GSO has further complicated the situation. The United States and most of the world have opposed the notion of sovereignty. The issue was addressed at the UNISPACE '82 Conference and its report stated: "Clearly, such a planning method should take into account the specific needs of the developing countries, as well as the special geographical situation of particular countries."[7] These ideas were subsequently adopted by the ITU Plenipotentiary in Nairobi although they were consistently opposed by the United States, other Western, and, indeed, Eastern Bloc countries.

Finally, and importantly, the LDC's have been and will continue to be the direct beneficiaries of hundreds of millions of dollars of R&D funds spent on satellite system design, which will lead to reduced costs for the systems they select to implement. Many of these funds are directed toward those technologies which lead to enhanced capacity. This capacity, however, cannot be enhanced if restrictive orbit planning methods are adopted.

Recommendations on Allocation of Geostationary Orbit Resources. The current regulatory scheme for obtaining access to the GSO is contained in the ITU Radio Regulations as revised at the 1979 WARC. It is based on accommodating orbit access on the basis of defined needs. Because of concerns of the developing countries that all available orbit locations may be used by the developed countries, however, there is growing concern in the United States that many LDCs will use the 1985/8 ITU Space WARC to establish a plan in which orbit and spectrum are pre-assigned rather than employed on the basis of need. It is the U.S. view that such an approach would be detrimental to space communications development and to the interest of all users.

The United States has supported and continues to support the concept of equitable access to the GSO. The United States, in preparing for the 1985/8 Space WARC[8] must develop proposals that promote flexibility and use of the orbit on an "as needed" basis. Most importantly, the United States must develop technical information and rationales that will assure the LDCs these procedures will provide "equitable access" to the orbit better than a pre-planned, long-term assignment approach.

Facilities Planning and Authorization

International satellite systems are planned within Intelsat, while cable systems landing in and used by the United States are planned jointly by U.S. carriers and their foreign correspondents. The Federal Communications

Commission (FCC) has exclusive authority to permit U.S. carriers' construction of new or additional domestic or international facilities. It also authorizes U.S. entities to provide basic services[9] over existing or new domestic and international facilities. The FCC, however, has no jurisdiction over the non-U.S. activities of foreign telecommunications entities. Nor, of course, can the FCC authorize U.S. carrier entry into, or provision of service in, foreign markets.

The planning and construction of new international facilities as well as the provision of international services requires a joint effort between U.S. carriers and their foreign correspondents, or between Comsat and the other members of Intelsat and Inmarsat. The FCC is placed "in the middle" of these joint activities. Moreover, the primary statutes[10] which authorize Commission action provide little guidance regarding the manner and extent to which national interest, foreign policy, national security, and international comity concerns must be balanced by the FCC in conjunction with its international facility and service authorization activities. While each of these concerns constitutes one component of the broad "public interest" standard under which the FCC operates, there are instances where it requested and followed Executive Branch advice on these subjects and other instances where such views were neither sought nor, when offered, accorded proper weight.

The ownership and management of satellites and cable facilities pose major differences in terms of their operations and capabilities, their economic characteristics, and planning. Submarine cables landing in the United States are owned and operated by the U.S. service carriers and their foreign correspondents, generally on a 50-50 basis. They provide point-to-point communications between two countries or continents, and are traditionally planned for and operated primarily under the control of those entities, in conjunction with any other administrations which purchase transmission capacity for extension to their domestic communications networks.[11] The total cost of a submarine cable system, with a typical design life of about 24 years, is established by including manufacturing and laying costs, operating and management costs, research and development, repairs, and other directly associated expenses. The costs of cable circuits have thus been relatively simple to determine.

International communications satellites are primarily planned, owned, and operated by the international organization Intelsat. Ownership of Intelsat is vested, according to use, in the various Signatories to the Intelsat Operating Agreement. Comsat, the U.S. Signatory, owns the largest share of the Organization. A single modern satellite can simultaneously accommodate many point-to-point or point-to-multipoint communications paths, thus providing great operational flexibility. Since earth stations accessing the Intelsat system are not owned by Intelsat, however, but by the individual administrations, significant international technical cooperation is necessary and has become routine. Furthermore, the cost of an Intelsat circuit is not as simply determined, since a major objective is the establishment of universal and flexible access to its global communications network. Earth station costs of each nation also are different. Finally, Intelsat prices reflect a significant amount of averaging among satellites and ocean regions, so that efficient high volume routes provide some subsidy to low volume routes. Economic and operational comparisons between satellites and cables thus have been, and are likely to remain, complex and controversial.

Current Policy Concerning Facilities Planning and Authorization. U.S. regulation of international communications is now accomplished by the FCC as required by the Communications Act of 1934 (as amended), the Communications Satellite Act of 1962, and the International Maritime Satellite Act of 1978.[12] With the exception of the amended Section 222[13], the regulatory scheme imposed by the 1934 Act does not distinguish between domestic and international common carrier facilities and services. Hence, the same broad "public interest" standard contained in the 1934 Act applies to all facilities and services, and there is no explicit legislative requirement for the FCC to take into account considerations such as the U.S. national interest or foreign correspondent requirements.[14] The Comsat Act also does not address the legal obligations imposed by the Intelsat Agreements, which were created after the Act was passed; neither the Interim nor Permanent Agreements were ratified by the Senate.

The FCC's public interest analysis regarding the licensing of international facilities and services has a major impact upon the plans, investments, and revenues of many foreign telecommunication entities. The FCC participates in the development of U.S. guidance to Comsat regarding Intelsat facilities and services. The FCC has also actively participated in the North Atlantic Consultative Process for the planning of both satellite and cable facilities although it lacks jurisdiction, obviously, over the activities of foreign telecommunication entities in these joint enterprises. It is difficult for the FCC to engage in meaningful unilateral review of international satellite facilities for which Comsat seeks authorization, when Intelsat has already approved the long-range planning for such facilities. This has led to greater FCC focus upon submarine cable facility applications, although, similarly, it also has proved difficult not to authorize cable facilities which are deemed necessary by U.S. service carriers and their foreign correspondents.

Because of the continuing problems with the FCC's application of outdated legislation, an alternative approach to the planning and authorization of international facilities was sought by a group of European Administrations. Starting with a 1974 meeting with U.S. Government representatives (including the FCC, the Office of Telecommunications Policy, NTIA's predecessor, and the Department of State), they sought to devise a procedure wherein cable and satellite planning for the North Atlantic region could include consultations with both U.S. carriers and Government representatives at an early stage. This procedure has evolved into a formal arrangement[15] now called the North Atlantic Consultative Process.[16]

By the late 1970s, the FCC processes for approval of international telecommunications facilities in general, and for Intelsat V and TAT-7 in particular, were the subject of much controversy in the United States as well as in Europe. Pursuant to a Congressional request, the General Accounting Office reviewed the overall situation and issued a final report in March, 1978.[17] The report made several cogent observations and recommendations, some of which have affected continued FCC activities and have been incorporated in proposed legislation.

There has been some attempt to extend the Consultative Process to the Pacific Ocean Basin and elsewhere. To date, the other nations of these areas have not indicated a high receptivity to the concept. The FCC has instituted a proceeding,[18] however, that attempts to involve joint cable/satellite planning for the Pacific Basin by U.S. carriers only. Whether such one-sided planning can be useful in the long term is unclear.

There are serious drawbacks in the current process which necessitate an improved long-term resolution of the international facilities and services planning and authorization process. These drawbacks include:

o Excessive economic control by the FCC over the international communications marketplace, evidenced by extensive allocation of cable and satellite market shares and detailed country-by-country approval of carrier plans;

o Undue FCC focus on engineering and economic factors, and too little consideration of national interest, foreign policy, and national security concerns; and

o The difficulties of addressing the facility and access needs of new and potential service providers when the cooperation of foreign correspondents is required, and when such new entrants seek to compete with those U.S. carriers owning existing facilities.

In summary, there is excessive economic regulation and it inhibits attainable market competition, diversity, and innovation. As a result, the existing process is inadequate to promote U.S. national interests.

Options concerning Facilities Planning and Organization. Resolving the problems in the planning and authorization of international facilities and services hinges on two questions. First, how can the advantages of full and fair competition and attendant deregulation be achieved, when there is one legislated monopolist, and one de facto monopolist, among U.S. carriers, and when the overseas correspondents of U.S. carriers are also monopolists? Second, in either the current U.S. regulatory regime or in a future, more competitive, deregulatory environment, how can the national interest, foreign policy, public interest, and national security be protected and promoted?

As these two questions have come to be understood and appreciated, alternatives to the current statutory process for international facility approval have been advanced. Before describing specific options to resolve the problems with the process as it is currently implemented, however, the detailed goals the U. S. should attempt to achieve in the planning, construction, and use of international facilities should be considered. These goals are:

- Access for U.S. telecommunications users to international facilities in sufficient diversity, quantity, and quality to ensure the low-cost, reliable choice of desired services.
- Minimal governmental regulation or oversight over U. S. carriers cooperating with foreign entities in the planning, construction, and operation of facilities.
- Encouragement of intermodal (cable vs. satellite) and inter-company competition.
- Statutory recognition of the growing importance of international facilities to U. S. national interests, foreign policy, and national security.
- Prohibition of the construction of inefficient or unnecessary facilities and other improper activities by "bottleneck" carriers in order to protect U.S. telecommunications users.
- Government recognition and amelioration of the disadvantages of competitive U. S. carriers attempting to negotiate with monopoly foreign correspondents while still accommodating the needs of international comity.

Some of the alternatives to the current statutory process for international facility approval, which have already been proposed as options by others include:

- Maintain the status quo; permit the Commission to continue the consultative process, and take other ad hoc actions as necessary.

- Make sections 214, 309, and 319 of the 1934 Act inapplicable to international facilities and services, thereby allowing facility choice to be determined by the carriers in response to marketplace forces and foreign regulatory actions.

- Create a Government/industry task force for the planning of international facilities.

- Require the FCC, or a joint U. S. Government group (e.g., FCC, State, Commerce/NTIA), to develop detailed guidelines, including cost comparison methodology for international underseas cable and satellite systems, and specific operational criteria, which would form the basis of Commission decisions on facility construction and operations.

- Retain section 214 substantially as is, but add an "alternative" international facility construction and authorization process permitting carriers to invest in and construct facilities without prior FCC approval, and with post-construction regulatory review. Carriers would then be free to select either avenue, weighing the benefits and risks of each.

Recommendation for Facilities Planning and Organization. None of the alternative approaches is perfect, nor will any single one solve all extant and foreseeable problems. Maintaining the status quo would be particularly troubling. The FCC has neither the legislative mandate nor the expertise to negotiate effectively with foreign governments. Furthermore, pending litigation could well require the FCC to discontinue the consultative process.[19] Existing legislation requires lengthy regulatory proceedings entailing costly uncertainties about projects involving hundreds of millions of dollars. Finally, the existing legislation permits the FCC to act without adequate consideration of the national interest or foreign policy of the United States at a time when these concerns are of increasing importance in international telecommunications.

INTEGRATED SERVICES DIGITAL NETWORK

National and international work toward an Integrated Services Digital Network (ISDN) is being carried out in various forums and countries around the world. Yet, ISDN is still only a concept. Many technical and standardization problems exist. Likewise, many domestic and international policy decisions must be made before ISDN comes into reality. As the network of the future, ISDN is a vital consideration in a study of long-range goals.

The ISDN Concept

The core concept of ISDN is a network that will be based on and evolve from the telephony integrated digital network (IDN) by progressively incorporating additional functions and network features, including those of any other dedicated networks, so as to provide for existing and new services (CCITT Recommendation G705). Telephone networks around the world have evolved through two distinct stages, and are embarking on the third. The first stage was completely analog, both transmission and switching, and was designed for voice transmission. The second stage began evolving with the introduction of digital transmission and switching. It is identified as IDN and, in addition to voice, carries services such as data and facsimile. The third stage will encompass end-to-end digital connectivity and provide the backbone of ISDN.

There are three factors motivating ISDN. They are: (1) new or expanded services which can be offered, (2) the economy or lower cost of offering the services because of digital network characteristics, and (3) new technology which can permit the new services to be offered at reasonable cost. Combining these factors results in economic benefits through services integration.

Since the present telephone networks already carry various services, there currently exists a form of integrated services network. And, thus, it is not the evolution and digitization of the network in itself that has made ISDN a major subject of national and international importance and a subject of debate. Rather it is the philosophy in the planning and design of the future network and of subscriber access to that network, including the interconnection of constituent (telephone and non-telephone) networks as well as peripheral networks (e.g., private networks).

Although all ISDN services have not yet been defined, or perhaps even envisaged, these services are expected to fall into the following categories:

- Digitized voice, with voice encryption a future possibility.
- Facsimile and graphics.
- Video. Whether digital TV will be provided is uncertain because of the large bandwidth required. Other video services are planned, however.
- Other services, including telemetry, videotex, software transfer, electronic mail, data base access, computers, and other terminals.

Because the ISDN concept is now only defined in very general terms, it is open to different opinions and interpretations. One reason for this is that some countries are better able to implement a total digital approach than others, because they do not have a large plant investment in the latest analog switch technology. Other reasons arise in the differences between countries in the degree of governmental control of telecommunication resources and their political and economic philosophy. The differences in interpretation of the ISDN concept are indicative of the nature of issues and standardization which will have to be addressed. Some examples of the conflicting views about the scope and impact of ISDN are:

- o There will be a single, worldwide ISDN.
- o Each country will have an ISDN with a (high) degree of international connectability.
- o Several ISDN's may exist within a single country with varying degrees of connectability.
- o The ISDN will operate in parallel with dedicated public networks and private networks and compete with them for customers.
- o All services will be provided by the ISDN with little room left for other telecommunication service providers and networks.

Foreign ISDN Influence

The drive toward establishing an ISDN has varying strength in different parts of the world. In the United States, the common carriers are adding interfaces to end-office analog switches to permit digital operation, while some countries, such as France, have essentially planned to leapfrog a technical generation by establishing a new digital telephone network. Many European countries and Japan have done major planning for ISDN and in some cases have or are establishing pilot and experimental networks which include not only metallic but also optical fiber and satellite transmission. A number of European countries seem serious about setting standards for ISDN soon. The United States should continue to participate strongly in the activities of international standard-setting bodies, so that the interests of our domestic manufacturers, service providers, and network operators are recognized, and so that U.S. procompetitive policies will be taken into account.

Domestic ISDN

In the United States, the emerging ISDN will evolve from the current and developing technology which includes digital switching, digital subscriber loops, digital transmission via lightwaves, satellites, digital data services and signalling systems.

The U.S. communications environment is characterized by a number of factors which do not, or only to a much lesser degree, exist in other countries.

Some of the major factors that may effect ISDN are:

- o Distinction between basic and enhanced services.
- o Multiplicity of domestic common carriers and, hence, public networks.
- o Multiplicity of private networks that need to be connected to the public network(s).
- o Development of alternative local distribution schemes.
- o Multiplicity of international carriers.

These factors raise critical policy and standardization questions. Standards in an analog environment are more forgiving than is true in an all-digital communications world. Ensuring full and fair access to technical, interface, and other standards as well as operating protocols is critical to ensure competition.

The distinction between basic and enhanced services will, from the U.S. perspective, have to be considered by the CCITT on the subjects of services to be provided by ISDN and on service classification.

U.S. Domestic Issues

There are numerous issues relating to domestic ISDN implementation. Three major issues which appear to subsume a number of others are described in this section.

Impact on Procompetitive Policy. The evolution of the ISDN raises serious questions as to some of the assumptions underlying the U.S. procompetitive policy to date. One viewpoint has been expressed as follows:

> It will be increasingly difficult to ground service distinctions in regulating network facilities. The multi-service and multi-facility (integrated packet and circuit switching) environment increasingly relying on or utilizing new network storage and processing (not necessarily "computing") resources will frustrate attempts to regulate the "basic-enhanced" service dichotomy and related segregated facility provisions.[34]

General Competitive Access in the United States. There are two types of access of concern with regard to ISDN: (1) access by service providers, and (2) access by equipment providers. In the case of multiple service providers, to what extent will the U.S. requirements of full interconnection be reflected in the U.S. ISDN thereby providing maximum competitive access? In addition, will the same level of interconnection apply to private networks, by-pass carriers, broadcasters, and information service providers?

Opportunities for suppliers of customer premises equipment (CPE) could be limited not only by normal interconnect-type barriers, but also by the unique complexity of ISDN. An example of this concern has been described as follows:

> The ISDN also greatly complicates the segregation of service markets from CPE markets when standards, functional allocations and performance parameters of the ISDN significantly affect a CPE required. The concern is that the network service provided could configure network equipment so as to favor his own CPE, or (by placing most of the "intelligence" in network equipment) to reduce the role of the CPE and thus the size of the CPE market.[35]

Multiple ISDNs. The two major concerns relating to the existence of multiple U.S. domestic ISDNs are: (1) will they be interoperable, particularly in a deregulated market, and (2) how will the interoperability or lack thereof affect national security requirements?

International and Global Considerations

In addition to the domestic issues, there are a number of more fundamental questions which arise in the international and global sphere:

> Underlying these policy issues are basic ideological and even philosophical questions. Will ISDNs be configured to allow major roles for competing private enterprises, or will the public service monopoly principle prevail? How are the inevitably divergent policies in different nations to be accommodated at the international level? Can countries favoring a dominant PTT role be expected (or entreated or pressured) to open their equipment and service markets to "free trade"? Will advocacy of a public sector monopoly approach justify technical standards-setting behavior which pro-competitive governments may regard as "protectionist"?
>
> Viewing the longer-term and the broader world beyond the major industrial countries, even larger issues may loom. Should ISDNs be configured to serve developing countries as efficiently and powerfully as developed countries (even if the economics are less favorable?) Could affordable access to ISDNs (and its productivity-enhancing services) significantly improve the position of some developing countries in the world "economic order"? Might unfavorable terms of developing country access exacerbate north/south differences over the future shape of the world "information order"?[36]

Recommended Actions. The effort to establish basic parameters for the ISDN, and access thereto, is seen by many administrations around the world as an urgent matter. That is because it is viewed as essential that standards be agreed upon before national implementations have reached a point where compromise will be too difficult to achieve.

Progress in developing ISDN standards is underway within the CCITT. U.S. Government and industry participants have devoted considerable effort and have contributed significantly in this undertaking. However, because of the far-reaching impact of ISDN on competition, equipment and service trade, and U.S. technological leadership, it is important that the U.S. Government provide a reasonably specific policy framework on which future U.S. efforts to develop ISDN standards could be based. Toward this end the FCC, which is presently considering a Notice of Inquiry on ISDN development, should initiate it without delay.

NOTES TO CHAPTER FIVE

[1]These microchip-based techniques are both alternative and complementary methods of increasing the efficiency of the basic transmission medium to transmit voice type information.

[2]Ibid.

[3]A GHz = gigahertz = 1,000,000,000 cycles per second.

[4]The ITU Allocation Table does not specifically refer to military applications although these are generally recognized.

[5]The ITU definition for Fixed-Satellite Service: A radiocommunication service between earth stations at specified fixed points when one or more satellites are used; the fixed-satellite service may also include feeder links for other space radiocommunication services.

[6]1979 WARC, Resolution 3.

[7]UNISPACE '82 Report.

[8]See discussion of ITU and Radio Conferences, infra.

[9]The FCC recently ruled that it would not regulate either domestic or international "enhanced" service providers as common carriers. See the following chapter for a full discussion of this ruling.

[10]Communications Act of 1934, 47 U.S.C § 151 et seq.; Communications Satellite Act of 1962, 47 U.S.C. § 701-744.

[11]In modern, high-capacity cables the ownership tends to become very diversified (e.g., TAT-7 has 18 European owners) and the destinations become more widespread (e.g., trans-Atlantic cables serve as transit facilities to Africa and Asia).

[12]A new Section 222 was inserted into the 1934 Act by the Record Carrier Competition Act of 1981.

[13]By contrast, Section 201(a)4 of the 1962 Act and Section 504(b) of the 1978 Act direct the President to ensure that the actions of Comsat are "consistent with the national interest and foreign policy of the United States." However, the President

does not have a statutory role in oversight of FCC decisions regarding Comsat or any other carrier, except in a wartime emergency. The President delegated his responsibilities under the 1962 Act to the Departments of State and Commerce. No delegation of Presidential responsibilities under the 1978 Act has been made.

[14]47 U.S.C. §751 et seq.

[15]See, e.g., Overseas Communications, 71 FCC 2d 71.

[16]Certain aspects of this process were challenged in ITT World Communications, Inc. v. FCC, No. 80-0428 (D.D.C. Oct. 17, 1980).

[17]Report of the Controller General, "Greater Coordination and a More Effective Policy Needed for International Telecommunications Facilities," CED 78-87, March 31, 1978.

[18]CC Docket 81-343.

[19]ITT v. FCC, D.D.C. Oct 17, 1980.

[20]Senate Committee on Commerce, Report on Communications Satellite Act of 1962, S.Rep.No. 1584, 87th Cong., 2d Sess. 1 (1962).

[21]Because the ownership of the satellite system was expected to be based upon a series of bilateral arrangements similar to underseas cable agreements, some of the provisions of the Act cite the system and the corporation interchangeably.

[22]Agreement Establishing Interim Arrangements, August 20, 1964, 15 UST 1705, TIAS No. 5646.

[23]Definitive Agreements for the International Telecommunications Satellite Organization, entered into force 12 February 1973, 23 UST 3813, TIAS No. 7532.

[24]"Satellite Communications," October 1971, U.K. Stationery Office, Cmnd. 4799.

[25]MARISAT, Inmarsat's predecessor, was a satellite project launched by Comsat for use by the U.S. Navy as well as commercial maritime interests.

[26]In at least one, and perhaps several instances, the capital investment for Intelsat space segment allotments and earth stations, as well as their operation, are not provided by the Signatory of the nation wherein the services are delivered.

[27]The Intelsat Agreements state:

- "Each State Party . . . shall designate a telecommunications entity . . . to sign the Operating Agreement. (Article IIb)
- "Relations between . . . . the Signatory and the Party . . . shall be governed by applicable domestic law. (Article IIb)
- "Each Signatory . . . to which an allotment [of space segment] has been made . . . shall be responsible for compliance with all the terms and conditions . . . unless . . . its designating party assumes such responsibility for allotments made with respect to all or some of the earth stations not owned or operated by such Signatory." (Article 15c)

[28]Whereas the 1962 Act describes and sets forth the structure of a single global satellite system, a clause added to Section 102(d) during final passage of the legislation at the behest of Senator Church suggests that additional systems could be desirable if the initial system were a failure or it were otherwise in the national interest.

[29]U.S. Department of State, Correspondence from the Under Secretary of State to the Chairman of the FCC, July 23, 1981.

[30]The U.S. Government is specifically noted as an authorized user in Section 305(b)(4) of the 1962 Act. On one previous occasion, the FCC had waived its restrictive policy. In 1978 (70 FCC 2d 2127), at the request of the Spanish International Network, the FCC authorized Comsat to provide Intelsat facilities at their earth stations directly to users of international television services. Several of the carriers appealed this decision in the courts.

[31]90 FCC 2d 1394 (1982).

[32]5 FCC2d 812 (1966).

[33]CC Docket 82-540.

[34]Martech Strategies, Inc., "ISDN — Integrated Services Digital Networks: Impacts and Industry Strategy," page 8-6, June 1982.

[35]Id. p. 8-6.

[36]Id. pg. 8-5.

Chapter Six

INTERNATIONAL TELECOMMUNICATIONS SERVICES

The provision of international telecommunications services is vital to the national interest of the United States, and the significance of these services in our economy is growing daily. This section summarizes current U.S. policy on competition in international telecommunications services, and analyzes resulting issues and their future consequences for U.S. interests.

BACKGROUND

Since the late 1960s, there has been a fundamental shift in the philosophy underlying U.S. telecommunications policy from pervasive economic regulation toward reliance on unregulated marketplace competition. This change has been soundly based on the premise that the benefits normally considered in the public interest -- a strong U.S. telecommunications industry with competitive and efficient rates, high quality of service, wide choice of services, and innovative technology -- can best be achieved through maximum possible reliance on competitive marketplace forces rather than through the regulatory processes of the Federal Communications Commission (FCC).

The FCC commenced the expansion of international telecommunications competition in December 1979.[1] Since then the FCC has taken a number of actions to increase competition in both the international voice and record markets. The international record carriers have been granted additional domestic and international authority. New firms have been authorized to provide international service and Western Union Telegraph was authorized to reenter the international market in accordance with the Record Carrier Competition Act of 1981. In December 1982, AT&T was permitted to enter the international record market and the record carriers to enter the voice market. The FCC has also undertaken a number of initiatives to restructure U.S. access to the Intelsat system. In August 1982, it completely removed all restrictions on noncarrier access to Comsat. Comsat was also permitted to enter the end-to-end service market with certain structural safeguards.

The relationship between regulation and competition is an integral part of all of the issues discussed in this section. A central question that must be addressed is what minimal amount of regulation or residual Government oversight is needed to achieve the goal of promoting the maximum possible amount of competition.

Before discussing specific issues associated with international telecommunications services, it would be helpful to clarify a few basic considerations, namely: (1) the institutional differences between domestic and international telecommunications; (2) trends regarding monopoly and competition in foreign countries; and (3) different levels of competition in the international environment.

Institutional Differences Between Domestic and International Telecommunications

Although international telecommunications service is a cooperative venture of the United States and other sovereign nations, not all the market opportunities existing in domestic telecommunications can be automatically achieved in the international arena. In a cooperative international venture, the United States cannot unilaterally mandate policies, structures, or market opportunities. While reliance on competitive market forces serves as the basis of communications policy in the United States, this is not generally the case overseas. In addition, safeguards intended to encourage fair competition, such as antitrust legislation, rate-of-return regulation of underlying facilities, and requirements for nondiscriminatory interconnection between carriers, are only applicable generally in the United States.

In most countries, telecommunications equipment, service offerings, rates, and conditions are controlled by a government-owned monopoly whose motivation and objectives are different from those of private firms in the United States. For example, in many countries, international telecommunications is run by the government as a profit-maximizing business, and international revenues are used to subsidize postal and other domestic services, including domestic telecommunications services. This is not the case in the United States, and the differences must be addressed in policymaking, or they will obstruct the implementation of U.S. international telecommunications objectives.

International carriers in the United States have not only operated under different economic conditions and motivations, but have also been subject to a unique regulatory scheme. In the past, the foreign administrations have adjusted to U.S. conditions because of the very large American traffic streams and the need for access to U.S. technology. In part because of the changes in U.S. regulatory policy, however, this stance is being reviewed.

U.S. efforts to apply a deregulated competitive approach in the international arena are a matter of some concern to several foreign telecommunication administrations. They have reacted in various ways. For example, some have declined to interconnect promptly with new U.S. carriers; and some have tried to exploit the new U.S. environment by having U.S. service providers bid against each other for the exclusive right to carry data services.

Monopoly and Competition in Foreign Countries

In July 1982, NTIA commissioned a study of 17 countries to determine the extent to which each was moving toward competition in the provision of telecommunications and information products and services.[2] Preliminary findings indicate that Japan has been considering a liberalization of its telecommunications marketplace and, indeed, plans to restructure Nippon Telephone and Telegraph Company, the domestic monopoly, into regional and long-distance enterprises (similar to the court-approved restructuring of AT&T) are now underway. Australia is also studying the possibility of adopting a more procompetitive policy, primarily in the domestic market. The Davidson Committee there, for example, recently proposed deregulation of customer premises terminal equipment and liberal entry criteria for private networks.

Canada and the United Kingdom have already begun moving toward certain limited forms of service competition. In the case of the United Kingdom, for example, British Telecom has been restructured (with the creation of a separate international subsidiary), a private specialized carrier, Mercury, has been authorized and, indeed, has recently been permitted direct access to Intelsat under liberal conditions. While British Telecom is likely to control a significant share of British telecommunications services for the foreseeable future, "privatising" the organization and fostering competition are major goals of the Thatcher government.

Most of the remaining countries in the study appear to favor the status quo. Some countries apparently wish to increase the role of the government monopoly. In those cases where there does exist some movement toward selective competition, the focus appears to be exclusively on the domestic market. Monopoly control of international service would be maintained. There is affirmative movement toward fostering a competitive environment in telecommunication services in some countries. Telecommunications deregulation has thus become in some senses a major American export especially given the positive experiences of the United States. In summary, however, the global environment is likely at best to be characterized by "selective competitive entry" (limited to certain service categories) and "administered competition," managed through government licensing and standards-approval authority for yet some time.

Levels of Competition

Within the United States, competition in telecommunications service generally involves intensifying commercial rivalry among U.S. carriers. This level of competition, however, cannot be automatically projected to the international arena, where two additional levels of competition must be considered.

In the international market, U.S. carriers and service providers must also deal with foreign telecommunications administrations (usually government-owned monopolies). The issues at stake in this international tug-of-war include the division of revenues between U.S. carriers and foreign administrations, the types of services authorized, the terms of interconnection, the technical characteristics of transmission facilities, and their place of manufacture.

The third level of competition occurs among the nations themselves -- the United States and foreign countries. The subjects of this level of competition include national security, employment, technological leadership, balance of trade, and influence in the determination of global telecommunications standards.

ISSUES

From the preceding background it should be apparent that some conflict exists between the desires of the United States to create a competitive market for international telecommunications services, and the perceived intention of some foreign administrations to maintain a monopoly market structure. How long this conflict will persist is unclear. The same technological forces chiefly responsible for competition in the United States are increasingly being felt abroad. If the implications of foreign monopoly power in telecommunications services are not fully recognized and adequately addressed, however, the United States may not be fully effective in its efforts to promote competition in these areas.

The issues discussed in this section include:

- The Resale and Computer II Decisions;
- International private leased service;

- o International operating agreements;
- o Financial arrangements for international switched services;
- o Foreign entry into U.S. international telecommunications service markets;
- o Competition in voice services; and
- o U.S. participation in CCI Standards Activity in a Deregulated Environment.

<u>The Resale and Computer II Decisions</u>

As mentioned earlier, the FCC started taking steps towards improving international telecommunications competition beginning in December 1979. The actions that have generated the most debate are: (1) the FCC's Notice of Proposed Rulemaking for International Resale and Shared Use of Services and Facilities, and (2) the FCC's Second Computer Inquiry (Computer II), which commenced in 1976. Because they relate to subsequent issues, these two actions are discussed more fully.

Resale is the subscription to communications services and facilities by one entity and the reoffering of services and facilities to the public (with or without adding value) for profit. Sharing is a nonprofit arrangement in which several customers collectively use communications services and facilities provided by a carrier, with each user paying the communications related costs according to its pro rata usage. In 1976 the FCC required carriers to remove from their interstate tariffs restrictions prohibiting the resale and shared use of private lines. In that decision the FCC concluded that resale and shared use of private line services would yield a number of public benefits, including:

- o the curtailment of unwarranted price discrimination (i.e., charging different customers different prices for essentially the same service) and possibly provision of communications services at rates more closely aligned to actual costs;
- o better management of communications networks, and facilitating the availability of noncarrier management expertise by intermediaries to users;
- o the avoidance of waste of communications capacity; and
- o the creation of additional incentives for research and development of ancillary devices to be used with transmission lines.

In 1980 the FCC came to a similar decision regarding resale of domestic MTS and WATS service.[3]

The question of international resale and shared use was addressed as part of the 1976 decision but the issue was deferred in a 1977 reconsideration.[4] Among the factors contributing to the FCC deferral was the international debate then underway. Resale and shared use had been subject to considerable debate in the International Telephone and Telegraph Consultative Committee (CCITT). In addition, many foreign administrations had expressed opposition to the initial FCC decision. The CCITT at its Sixth Plenary session in 1976 adopted its D.1 Recommendation prohibiting resale and shared use of international private lines and the United States concurred in this Recommendation. Because the issue was pending at the FCC, however, the United States added a footnote to the D.1 Recommendation, stating: "There is in the USA, a continuing discussion of issues related to the dedicated use of customer private leased circuits, and the U.S. delegation is pleased that this will be the subject of further study in Study Group III during this study period."[5] The footnote was intended to notify other CCITT members that the issue was under review in the United States and that we expected the Recommendation to be reviewed by the CCITT in its next plenary period (1977-1980).

In April 1980, the FCC again raised the subject of international resale, adopting a Notice of Proposed Rulemaking "to consider whether or to what extent common carriers subject to our jurisdiction which provide international communications services should be allowed to continue to restrict the resale and shared use of the services and facilities they offer under tariffs filed with the Commission."[6] The responses from U.S. parties divided along predictable lines. The established international record carriers, who benefit from the present resale restrictions, opposed the change. AT&T took the position that benefits could result from resale and sharing if the policy is carried out under competitive conditions. A large number of users also filed in the proceeding. In general, users supported the concept, although many recognized it was a controversial international issue and the larger users, especially, feared retaliation.

The benefits that would flow from resale and shared use include opening the possibility for a wider variety of services, promoting more efficient use of network facilities, and reducing the costs and potentially the prices charged users. Under the current CCITT restrictions, only large users can obtain the cost saving benefits of flat rate private lines. While the costs savings of resale and shared use could stimulate demand, foreign administrations also see it diverting traffic from the higher priced switched services. The additional revenues flowing from increased demand could offset any revenue losses associated with traffic diverted from switched to private line service. Some foreign administrations nevertheless fear that, on balance, they would lose.

The CCITT D.1 Recommendation was one agreed upon mechanism for preventing this diversion. Naturally, foreign administrations are concerned that, if the United States were to permit unlimited international resale and shared use, it would be difficult for them to prevent such activities, although they control one end of the facility. As a further measure, some administrations have threatened to retaliate by imposing volume-sensitive pricing for private lines or eliminating them altogether. Either action could greatly increase the annual cost for large users of international telecommunications such as DOD and U.S. companies that use leased channels throughout the world for their internal communications. It could also have a major effect on U.S. data processing firms providing services that compete with foreign companies, and in some cases with foreign administrations as well. In addition, possible small users' gains from resale services might be reduced or eliminated if volume-sensitive pricing were initiated.

The FCC's Computer II Rulemaking. Both the First Computer Inquiry (Computer I) and the Second Computer Inquiry (Computer II) represent efforts by the FCC to redefine and limit the scope of traditional common carrier regulation, taking into account the convergence of computer and telecommunications technology.

In Computer I, the FCC drew a distinction between communications services and data processing services, and concluded that the latter were not subject to common carrier regulation under Title II of the Communications Act. This determination applied to all domestic and international services provided over common carrier facilities. As the technology advanced, however, the criteria adopted in Computer I for distinguishing between "communications" and "data processing" services became increasingly obsolete. Accordingly, the FCC adopted new criteria for determining the scope of its Title II jurisdiction, based on a distinction between "basic" and "enhanced" services. Under Computer II, providers of basic services are regulated as communications common carriers, but enhanced services fall outside the scope of the FCC's Title II jurisdiction. The FCC defined the dichotomy as follows:

> We find that basic service is limited to the common carrier offering of transmission capacity for the movement of information, whereas enhanced service combines basic service with computer processing applications that act on the format, content, code, protocol or similar aspects of the subscriber's transmitted information, or provide the subscriber additional, different, or

restructured information, or involves subscriber interaction with stored information.[7]

In Computer I the FCC decided not to regulate data processing services; in Computer II, it concluded that enhanced services (which encompass both data processing and some transmission component) should not be regulated. The FCC's action in Computer II has thus served to reduce further the scope of common carrier regulation in an increasingly competitive environment. In its reconsideration of the final Computer II order, the FCC indicated that the basic/enhanced dichotomy applies to international services as well as to domestic services.[8]

While a broad variety of new services would be deregulated, Computer II does not deregulate basic transmission service nor in any major way alter the FCC's Title II jurisdiction over the common carrier transmission facilities used to provide international services. The FCC structured Computer II so that carriers authorized to own international transmission facilities must provide basic services under traditional common carrier regulation. An international carrier must therefore separate or unbundle its enhanced services from its basic services so that other enhanced service vendors can use the basic transmission facilities in a manner consistent with the carrier's tariffs governing use of the facilities.

The Commission adopted its Computer II decision in April 1980. Parties have subsequently challenged a Commission determination that the basic/enhanced dichotomy applies to international services. As recently as August 1982, the FCC affirmed that Computer II applies to international services.[9] Petitions were filed seeking reconsideration of the FCC's August 1982 Order.[10] Some petitioners contend the FCC has taken precipitous action without adequately considering the international ramifications. They argue, for example, that the FCC has, in effect, already ordered international resale and thereby prejudiced the outcome of its International Resale proceeding. If this were true, there would be cause for concern over possible foreign reactions.

Insofar as resale is concerned, the FCC has made clear that Computer II in no way alters existing prohibitions against international resale. When confronted with this issue, the FCC in its August 1982 Order stated that Computer II does not "prejudge issues pending before the Commission in International Resale." Order at para. 22. This is supported by the fact that the FCC has not required carriers to alter any tariff provisions restricting third party use that were in effect prior to Computer II. In addition, Chairman Fowler in a response dated October 20, 1982, to correspondence from Mr. Burtz, Director of the CCITT, stated that "[a] ny interpretation of Computer II to the effect that the Commission has established a new policy with respect to international resale or otherwise prejudiced the outcome of the International Resale Proceeding is simply incorrect."

It appears that some misunderstanding may exist among some foreign administrations as to the effect of the FCC's Computer II decision. The cause for this uncertainty is unclear. It is apparent, however, that some of the misunderstanding can be attributed to actions taken by certain U.S. carriers. Western Union International (WUI), for example, has appended to its Petition for Reconsideration of the FCC August 1982 Order correspondence from various foreign administrations expressing concern over the Computer II decision.[11] An examination of WUI's outgoing correspondence to these foreign administrations reveals, however, that concern on their part may be premised on misunderstandings as to what the FCC did in Computer II. Statements are made in WUI's correspondence that "the FCC has held that 'enhanced services' (i.e., value added services) may be shared and resold without limitation." WUI goes on to state it concerns that "this trend might broaden the boundaries of enhanced services to include basic services such as telex, message telephone, and leased channels . . . ."

These statements appear inconsistent with past statements of the FCC and inconsistent with the underlying theory of Computer II -- that basic services would continue to be regulated as common carrier services. In addition, the Commission has stated that "application of the basic/enhanced dichotomy to the international arena will not alter existing operational relationships between enhanced service providers and their foreign correspondents. Enhanced service providers will have to adhere to established practices with respect to offering international services." (August 1982 Order at n.7.) It appears, therefore, that some misunderstanding currently exists among foreign administrations as to the scope of FCC's Computer II decision. Any misunderstanding should be clarified so that it does not needlessly precipitate actions by foreign administrations that are adverse to U.S. interests.

Issues of major concern to the U.S. are (1) the continued availability of flat rate leased circuits, (2) the possibility some foreign monopoly administrations may exploit financial arrangements to the disadvantage of U.S. industry, and (3) the potential for unfair competition resulting from unregulated foreign-owned U.S. service providers acting in concert with their overseas affiliate. While these issues will be discussed in the following sections, certain observations are warranted here. First, application of the Computer II basic/enhanced dichotomy to international services is sound public policy. There is no need to subject enhanced services provided over common carrier facilities to a common carrier scheme of regulation. Experience gained over the last decade as a result of Computer I demonstrates this. Second, even though individual enhanced services are not subject to traditional common carrier regulation, regulatory action may be required to prevent foreign entities from utilizing U.S. transmission facilities to the competitive disadvantage of U.S. service vendors.

Whether the FCC can provide effective regulatory safeguards sufficient to protect U.S. interests in an unregulated environment remains to be seen. Under Computer II the FCC retains a full panopoly of regulatory authorities to license common carrier facilities and regulate their use. The court of appeals has affirmed the FCC's ancillary authority with respect to enhanced services where necessary to effectuate the FCC statutory mandate. AT&T v. FCC, 572 F.2d 17 (2d Cir. 1978). How the FCC will exercise its jurisdiction to safeguard U.S. international communications interests is unclear. The FCC appears to be proceeding on the assumption that so long as it retains jurisdiction to regulate the common carrier transmission facilities, over which enhanced services are provided, it will be able adequately to meet its responsibilities in accord with its statutory mandate.

Current Policy. Restrictions on third party use are currently contained in the tariffs of various international carriers. These restrictions serve to prohibit resale of private line services. The FCC is currently examining whether these resale restrictions should be removed.

The FCC has deregulated various services provided internationally. Computer II removes the provision of international enhanced services from the scope of common carrier regulation under Title II of the Communications Act. While the FCC has removed regulatory barriers to entry, an enhanced service may be offered internationally only if it is provided consistent with tariff provisions governing the use of common carrier transmission facilities and meets the approval of foreign administrations.

International Private Leased Service

Private leased service[12] is one of three basic network options — public switched, packet switched, and private leased networks. Large users, both U.S. and foreign, make extensive use of all three options in order to satisfy different technical needs. Private leased service is especially important to maintain highly-sophisticated business networks, because of the following unique advantages:

(1) Private lines permit the user to attach customer premises equipment of the latest technology, thereby maximizing circuit utilization. Private lines thus encourage innovative technology.

(2) Private line circuits offer more efficient and technically superior services, including quicker response time and a broader range of applications.

(3) Private lease offers substantial cost savings for a user with sufficient traffic to cover the relatively high fixed costs involved.

The continuation of international private leased service is a seldom-recognized but important U.S. concern. In part because of the controversy over application of the FCC's Computer II and international resale actions, some overseas administrations may review their policies concerning private leased service. Should this happen, the U.S. business community and the Department of Defense would face major cost increases for international telecommunications, estimated in DOD's case to be as high as 300-700 percent.

Private leased service is technically feasible even in an Integrated Services Digital Network (ISDN). Not all foreign administrations that favor ISDN are opposed to private leased service. In fact, there are distinct economic, technical, and political advantages to retention of private leased service by other countries. In order to retain this type of service within an ISDN environment, however, the U.S. Government must promote it vigorously, while encouraging broader participation in the ISDN development process.

International Operating Agreements

One impediment to increased entry of U.S. service providers into international telecommunications markets has been the reluctance of foreign administrations to sign operating agreements with additional U.S. carriers. In most cases, an operating agreement must be negotiated by a U.S. carrier and foreign administration prior to the initiation of direct telecommunications service between countries. The operating agreement usually sets all relevant financial and operating arrangements.

Most foreign administrations appear reluctant to operate with a large number of carriers, because there are added administrative and operating costs associated with each new carrier. Although not all European markets may be large enough to support an unlimited number of U.S. carriers, recent actions by some European administrations have nonetheless caused concern in the United States. The Nordic and Benelux countries recently indicated interest in possibly limiting the number of U.S. carriers for public data communications service between those countries and the United States, on the basis of competitive bids.

Even prior to deregulation, U.S. policy favored multiple suppliers in the record communications marketplace. The Nordic-Benelux proposals were originally perceived as a step backwards from the status quo, because they implied an interest in reducing the number of carriers of a particular service to only one -- creating single carrier exclusivity.

The positions of the Nordic and Benelux administrations were clarified after they became aware of several U.S. concerns. Informal discussions between representatives of the U.S. Government and the Nordic and Benelux administrations were held in Paris during January 1983. These discussions have led the U.S. to interpret each inquiry in the following manner:

> — The inquiry should not be understood as a formal request for competitive bids, but as a commercial inquiry seeking information about new services previously unavailable;
>
> — It is, similarly, not intended to lead to renegotiation of existing arrangements, but is a preliminary inquiry about interconnection with U.S. carriers for the new services;
>
> — Interconnection with multiple (four, five or more) U.S. carriers for new services is envisaged;
>
> — There is no intention to alter current arrangements in respect to accounting rates and their division.[13]

Current U.S. policy is to encourage an environment in which users are free to choose among competing suppliers on the basis of price and service quality. To this end, the United States favors an international market in which new carriers are free to enter, and long-term exclusive arrangements between foreign administrations and U.S. firms are not permitted.

Financial Arrangements for International Switched Services

The mechanism used to divide the revenues earned by jointly-provided international telecommunications switched services between U.S. carriers and foreign administrations is known as the accounting and settlement process.[14] Unfortunately, without U.S. Government oversight, this process could be used by foreign administrations to "whipsaw" one U.S. carrier against another. The current policy of the FCC is intended to help prevent such exploitation.

U.S. carriers generally share revenues from the switched services with their foreign correspondents, based on a negotiated "accounting rate." This rate is distinguished from the "collection rate" (the rate the customer pays) at either terminal. During a given month, if the number of paid minutes from calls billed in both countries is equal, the U.S. carrier and its foreign correspondent would be entitled to the same amount of accounting rate revenues, and no transfer of funds would be required. If, however, the number of paid minutes is greater in one direction than the other, the carrier with the higher outbound paid minutes must make a payment to its correspondent.[15]

In 1980, for example, there were more inbound minutes (to the United States) of both telex and packet switched services than outbound, and more outbound (from the United States) telephone minutes and telegraph words than inbound. Therefore, the U.S. carrier industry as a whole was in a receivable or creditor position with respect to telex and packet switched services, and was in a payable or debtor position with respect to message telephone and telegraph services.

The term "whipsaw" has been defined by the FCC as "the ability of the foreign correspondent to utilize its monopoly power to play one carrier against others to gain concessions and benefits from the U.S. international carriers."[16] In the past, only international record carriers (IRCs) have been deemed vulnerable to "whipsawing"; switched international voice services were supplied by just one carrier, AT&T, in the continental United States. The FCC authorized the IRCs to provide voice services in its decision of December, 1982. Carriers such as Comsat, MCI, and others may also provide voice services if they can obtain the necessary operating agreements with foreign administrations.

In the past, the foreign administrations have had the potential power to "whipsaw" the IRCs for two reasons: (1) the foreign administrations are typically monopolists, with the power to choose which U.S. carriers they will have an operating agreement with; and (2) these administrations generally control the distribution of return traffic to the United States among those carriers with whom they have signed operating agreements.

As an example of "whipsawing," a foreign administration might insist that a new telex carrier agree to a telex accounting rate lower than the prevailing rate in order to be allowed to interconnect with that country. Alternatively, an established carrier might be threatened with a reduction in its share of return traffic if it did not agree to a lower telex accounting rate. Once the accounting rate for one carrier is lowered, other U.S. carriers would have to agree to reduced accounting rates, to avoid a loss of inbound telex traffic.

The process can then be repeated by the foreign administration, starting with a different carrier. In this fashion, one carrier can be played off against the others, with the foreign administration constantly gaining a larger share of total revenues. "Whipsawing" thus diverts revenues from the U.S. industry to the foreign administrations, and reduces the potential for any significant reduction in U.S. users' rates.

"Whipsawing" could actually result in increased rates for U.S. users. On one hand, the U.S. carriers may simply "swallow" their loss, with no resultant compensatory rise in the U.S. collection rate. On the other hand, the carriers may have to compensate for the revenue loss by raising the U.S. collection rate. In either event, however, the U.S. balance of payments is adversely affected.

The marketplace cannot adequately protect the U.S. industry from the monopolistic tactics of foreign administrations in the accounting and settlement process. Once it became evident that "whipsawing" could occur, some U.S. Government intervention was necessary to protect the U.S. industry. The Government agency most deeply involved in the process has been the Federal Communications Commission (FCC).

To prevent "whipsawing," the FCC in 1980 reaffirmed a long-standing policy requiring uniform settlement rate agreements on parallel international routes.[17] This means that all carriers must abide by the same accounting and settlement arrangements for the same destination. However, the policy as reaffirmed by the Commission allows some departures from uniformity. The Commission has stated:

> We will consider, on a case-by-case basis, waivers of the policy of uniformity. Our consideration of a waiver request will be in accordance with the public interest standard and will include examination of the potential effects of a grant on the ratepayer, the carrier, and the industry.[18]

This uniform settlement policy prevents "whipsawing" by foreign administrations, while allowing for waivers if such action is demonstrably in the public interest. Such a policy strengthens the industry, promotes lower consumer rates, and contributes to a favorable U.S. balance of payments in telecommunication services. Under Computer II as presently interpreted, however, the uniform accounting and settlement policy would not apply to enhanced services.

Foreign Entry into U.S. International Telecommunications Service Markets

In recent years, as U.S. deregulation has continued, the U.S.

telecommunications service markets have become increasingly attractive to foreign entities. A number of U.S. carriers and Government agencies, as well as Congress, have expressed concern about the openness of U.S. service markets in contrast to the continuing lack of such openness abroad. In this situation, with the lack of truly reciprocal overseas markets, there may be a need to determine to what extent foreign entry into U.S. international telecommunication markets could result in unfair competition with U.S.-owned firms. This is an issue Congress, in any event, is likely to address.

In general, the United States has sought mutually open international markets in services as well as equipment, as a matter of sound trade policy. Open world markets offer the consumer lower prices, better quality, and greater diversity; new technologies may be more rapidly exploited; exports are stimulated; and worldwide employment opportunities are expanded. In such a mutually open environment, foreign entry into U.S. domestic and international telecommunications service markets would be highly beneficial. In services as in product markets, however, it is important to assure that competition is fair both in U.S. and foreign markets.

In the product area, if a foreign company engages in "dumping" — flooding the U.S. market with large amounts of an item at an unjustifiably low price —U.S. companies can take legal action. Equivalent legal protections do not exist for services. While consumers may benefit in the short run from "dumped" services, the practice could have serious long term effects if it weakens U.S. industry and endangers the success of procompetitive policies. The United States should therefore be prepared to act against unfair competition in U.S. markets by foreign-owned telecommunications service providers.

There is always some potential for anticompetitive conduct implicit in a market in which numerous firms are competing vigorously. When a telecommunications firm operating in the United States is owned by a foreign government,[19] additional competitive issues arise:

(1) The foreign-owned firm may be able to benefit from discriminatory access to markets in the foreign country, through the parent foreign administration's monopoly control of the foreign half of international transmission facilities and networks.

(2) Foreign-owned U.S. firms may be able to benefit from more favorable financial arrangements with the parent country than are afforded U.S. firms.

(3) The foreign country may be willing to provide the foreign-owned firm with service between the United States and third countries at a lower cost than is offered to U.S.-owned firms (for example, service from the U.S. to Singapore, via Hong Kong).

(4) The foreign-owned firm may be able to benefit from a disproportionate share of return traffic from the parent country or countries.

Some of the practices described above could have some distortional effects and might be difficult to detect. Unfair competition by foreign-owned firms could be detrimental to the interests of the United States, and could have adverse effects on trade and national security.

The FCC has deregulated domestic resale carriers. There are statutory restrictions on foreign ownership of firms holding radio frequency licenses (47 U.S.C. Sec 310(b)). Foreign entities, however, can obtain transmission capacity from common carriers to provide basic or enhanced services. The Commission's recent Authorized User decision allows noncarriers to obtain transmission capacity from Comsat. Since the Commission did not restrict the types of entities that could obtain satellite circuits directly from Comsat, foreign entities can obtain, on an end-to-end basis, an international satellite circuit for private or commercial use. A foreign firm could operate in conjunction with a foreign administration in a discriminatory manner to the detriment of U.S. service vendors.

Anticompetitive activities by a domestic firm that have an impact on U.S. commerce are subject to the antitrust laws, whether the firm is U.S. or foreign owned. The fact of non-U.S. ownership confers no immunity from possibly severe antitrust sanctions. At present antidumping laws do not apply to the telecommunications services market. There are no explicit statutory requirements governing disclosure of financing, ownership, and other relevant information that are unique to foreign-owned firms. Authority to investigate anticompetitive activity by foreign-owned or other providers of enhanced services would reside in the Commission's ancillary jurisdiction respecting competition in commerce (e.g., 47 U.S.C. Sec. 313, 314). The stated policy of the U.S. Government has been to sanction foreign investment in this country and participation in our economy by foreign firms. The United States thus does not oppose the involvement of foreign-owned entities in our competitive telecommunications marketplace. At the same time, the potential competitive issues posed by such involvement that may be unique to telecommunications should be recognized.

Competition for Voice Services

During 1981, international record services in the U.S. accounted for only 19 percent of total international telecommunications revenues, whereas international voice service revenues were responsible for 81 percent[20] of the total. Congress and the FCC in recent years, however, have concentrated on increasing record service competition, and have paid comparatively little attention to competitive conditions in the voice market.

The Record Carrier Competition Act (RCCA) of 1981 amended Section 222 of the Communications Act of 1934, in order to improve domestic and international telecommunications competition in record service.[21] The new Act allows the Western Union Telegraph Company to compete in the international arena, and the IRCs to offer domestic telecommunications, without prior FCC authorization. It also requires nondiscriminatory interconnection among all record carriers. Recently, interpreting the general requirements of the RCCA, the FCC (a) required record carriers to unbundle their overseas transmission charges into separate domestic and international components, (b) established provisions to determine the charges for domestic handling of international traffic, and (c) established a formula to negate any advantage an IRC may derive from an operating agreement with a country that refuses to grant one to another U.S. carrier. The FCC has also released a Report and Order which eliminated the voice/record dichotomy in the provision of international service.[22]

Interconnection of voice services may be addressed by the FCC in the future. To ensure fair competition in the voice market, the FCC should, at a minimum, take steps similar to those adopted in the record communications business concerning interconnection. Unbundling of overseas transmission charges for voice services into separate domestic and international components as presently required for record services may be also necessary. These and other approaches to foster competition need to be evaluated.

U.S. Participation in CCI Standards Activities in a Deregulated Environment

International consensus on technical and operating standards and tariff principles is reached in the International Telecommunication Union (ITU). The results are incorporated in ITU regulations and in the recommendations of the International Radio Consultative Committee (CCIR) and the International Telephone and Telegraph Consultative Committee (CCITT).

United States participation in ITU activities is ultimately the responsibility of the Department of State. U.S. working groups of the CCIR and CCITT include representatives from industry, users, and various agencies of the Federal Government. The delegations are always headed by a U.S. Government representative.

In U.S. CCITT working groups, where preparations are made for international meetings, the Federal Government must act as organizer, mediator, and catalyst in arriving at consensus views among industry representatives. In U.S. CCIR preparatory groups, by contrast, industry plays a different role because the Federal Government has a direct responsibility for most matters covered in the CCIR. Although members of U.S. industry submit company papers and make presentations at international meetings, particularly CCITT meetings, only Federal officials can state the U.S. position. (There have recently been proposals to modify this restriction.)

The recommendations of these international committees are legally voluntary but are generally accepted throughout the world. Those in which the United States has concurred are the functional equivalent of international agreements. The U.S. Government ensures compliance with these international recommendations within the United States, principally by means of the authority granted the FCC under the Communications Act of 1934. The FCC powers consist primarily of tariff and certificate authority governing new carrier facilities and services (Title II) and radio licensing authority (Title III). In addition, radio frequencies for U.S. Government use are controlled and assigned by NTIA pursuant to section 305 of the 1934 Act and under authority granted it by Executive Order 12046. NTIA's responsibilities in this regard are carried out in accordance with ITU rules and recommendations.

Recommendations

Some Government involvement in the international telecommunications arena is required to protect vital U.S. interests. Full and fair competition cannot be ensured without Government oversight, as long as foreign administrations resist movement toward a truly competitive international marketplace.

At present, the United States is promoting a competitive marketplace while much of the world still lags behind. Other governments have taken commendable initiatives to foster competition, although more needs to be done.

The United States should continue aggressively to seek a more competitive international communications marketplace. We must make clearer to other nations that competition does not constitute a threat to their sovereignty and that the benefits of competition are substantial. We should also indicate that the same technological forces driving competition in the United States are having similar effects abroad. A possible approach would be for the United States to have more high level discussions with those segments of foreign governments concerned with foreign policy and economic matters. These policymakers would have less of a

vested interest in the existing telecommunications arrangements of their countries and could see the broader benefits of competition to users, consumers, and equipment suppliers.

NOTES TO CHAPTER SIX

[1]On December 12, 1979, the Commission took action in the following cases related to competition in international services:

(1) International Audit and Study, 75 FCC 2d 726 (1980);

(2) International Gateways, 76 FCC 2d 115 (1980);

(3) International DATAPHONE, 75 FCC 2d 682 (1980);

(4) Alternate Voice/Data Service and Interconnection with Domestic Network, 76 FCC 2d 166 (1980);

(5) Interface of International and Domestic Telex and TWX Services, 76 FCC 2d 61 (1980);

(6) New Telex Service Arrangement via Mexico and Canada, 75 FCC 2d 461 (1980);

(7) Consortium Communications International, Inc., 76 FCC 2d 15 (1980); and

(8) Regulation of Domestic Public Message Service, 75 FCC 2d 345 (1980)

[2]The NTIA study focused on Australia, Austria, Belgium, Brazil, Canada, France, West Germany, Hong Kong, Italy, Japan, Mexico, Philippines, Singapore, Sweden, Switzerland, the United Kingdom, and Venezuela.

[3]Resale and Shared Use of Common Carrier Services and Facilities, 60 FCC 2d 261 (1976); MTS/WATS Resale & Sharing, 80 F.C.C. 2d 54 (1980).

[4]Resale and Shared Use of Common Carrier Services and Facilities, 62 FCC 2d 588, 593 (1977) (reconsideration).

[5]CCITT Orange Book, Vol. II.1, "General Tariff Principles, Lease of Circuits for Private Services," p. 83 (VIth Plenary Assembly, Geneva 1977).

[6]International Resale, 77 FCC 2d 840 (1980) (notice).

[7]Amendment of Section 64.702 of the Commission's Rules and Regulations (Final Decision), 77 FCC 2d 384, (1980).

[8]Amendment of Section 64.702 of the Commission's Rules and Regulations (Reconsideration), 84 FCC 2d 50, 53 fn 4 (1980).

[9]FCC Memorandum Opinion and Order, "GTE Telenet Communications Corporation and Tymnet, Inc." Authority to extend Packet-Switched Services to

Western Europe, released August 25, 1982.

[10]Petitions for reconsideration of the FCC Memorandum Opinion and Order of August 25, 1982, were filed by WUI, RCA Global Communications Inc., Association of Data Processing Organizations, Inc., and International Communications Association. These petitions were supported by Control Data Corporation and GTE Corporation. Aeronautical Radio, Inc., and IBM supported having Computer II apply to the international arena.

[11]Western Union International, Inc., Petition for Reconsideration in the matter of GTE Telenet Communications Corporation and in the matter of Tymnet, Inc., Appendix A, September 24, 1982.

[12]Private leased line service in this chapter refers to the use of voice grade circuit bandwidths and greater.

[13]Letters from Matthew V. Scocozza, Deputy Assistant Secretary for Transportation and Telecommunications, Department of State, to Nordic and Benelux Administrations, January 28, 1983.

[14]Leased channel services do not entail such a process. Each administration charges a monthly rate for its half-circuit (from the country to the theoretical midpoint of the path.)

[15]The description of the process has been simplified somewhat for purposes of explanation.

[16]Uniform Settlement Rates on Parallel International Communications Routes, 84 FCC 2d 121, 122 n. 3 (1980).

[17]Uniform Settlement Rates, 66 FCC 2d 359 (1977) (notice); 84 FCC 2d 121 (1980) (order).

[18]84 FCC 2d at 128.

[19]For example, Pacnet Communications Corporation (PACNET), a resale common carrier, is a wholly-owned subsidiary of Cable and Wireless PLC, of which the British government owns 49 percent. Cable and Wireless and its affiliates serve as the Recognized Private Operating Agency (RPOA) for a large number of countries around the world.

[20]Includes AT&T and four other voice carriers. See Federal Communications Commission, Statistics of Communications Common Carriers, Tables 14 and 25 (1981).

[21]Pub. L. No. 97-130 (Dec. 29, 1981); 47 U.S.C. § 222.

[22]FCC Report and Order, Overseas Communications Services, CC Docket No. 80-632, released December 22, 1982.

Chapter Nine

RESEARCH AND DEVELOPMENT IN TELECOMMUNICATIONS AND INFORMATION INDUSTRIES

Longstanding U.S. leadership in telecommunications and information technology is being challenged by increasingly strong and concentrated foreign competition. Further erosion of the U.S. position has serious implications for the long-term competitiveness of the U.S. telecommunications and information industries and would also adversely affect the entire U.S. industrial sector. A thorough review is thus needed of current U.S. policy in research and development (R&D) as it relates to telecommunications and information.

This chapter discusses some of the factors that require attention. The material reviewed does not indicate the United States has neglected R&D funding, or that U.S. research and development no longer generates patentable inventions or advanced products. The evidence does indicate, however, that the United States no longer enjoys an overwhelmingly dominant position in "high-technology" industries.

BACKGROUND

For decades the United States enjoyed a position as a world leader in technology. Underlying U.S. technological leadership was its ability to generate new scientific knowledge and innovations, and to continually transform them into useful products and services in ways superior to foreign competitors. Investments in research and development have contributed importantly to exports,[1] job creation,[2] new products, productivity improvements[3], enhanced the quality of life, and accelerated advancements in science.

Projections for the 1980s indicate very rapid growth in high technology American industries. Projected growth rates include: for robotics (25 percent), computers (18 percent), semiconductors (12 percent), guided missiles and space vehicles (16 percent), and communications equipment (5 percent).[4] Each of these "sunrise" industries is highly dependent on telecommunications and information technology. Technological advances in these industries have brought improvements in productivity to all sectors of the economy, through such innovations as digital communications, electronic funds transfer, electronic mail, automated office equipment, factory assembly operations, and smaller, more affordable, computers for home, business, and scientific purposes.

Recent foreign technological advances create concern that the U.S. leadership position in high technology is eroding. U.S. trade deficits (especially with Japan), declining world market shares, increased competition for the U.S. domestic and international markets, and the rapid appearance of sophisticated foreign products, such as semiconductors, satellite communications, lasers, computers, consumer products, and robotics, are cited as evidence of a deteriorating U.S. position. The ability of other nations to select technologies for concentrated R&D focus, to provide government subsidies, to encourage joint government-industry cooperation, to devote resources to nonmilitary enterprise, and to protect indigenous enterprises from competition through tariff and nontariff barriers, are seen as factors contributing to foreign advantages.

A review of U.S. R&D in telecommunications and information technology has particular importance because technological leadership in these and related areas is widely seen as the cornerstone of economic advancement in the decade of the 1980s.[5] Nations having outstanding capabilities for generating and applying advanced technologies are likely to flourish.

This chapter reviews U.S. R&D in telecommunications and information technologies from a number of vantage points, and recommends improvements. Although substantial progress has already been made, continued efforts are needed to improve incentives for making the best use of U.S. genius for innovation and commercialization of new products. Aggressive foreign competition for world leadership and in telecommunications and information technology, as well as in other technological fields, is likely to be permanent. In this environment, the United States will have to use every possible means of maintaining a strong position capable of countering foreign policies.

R&D Perspective: Current Strengths, Trends and Problems

Government Role in R&D

Federal Government support for R&D activities falls into three categories: first, backing those activities which meet needs in which the Government is the sole or primary user (such as national defense); second, backing those activities which assure the strength of the economy and the welfare of its citizens (such as agriculture, energy, and health); and third, funding basic research. The 1983 Budget of the United States helps clarify the distinction between Government and private sector responsibilities for R&D. The Government views its role as follows:

- o providing a climate for technological innovation which encourages private sector R&D investment;
- o focusing its direct R&D support in areas where there is likely to be significant economic gain to the nation, but where the private sector is unlikely to invest adequately because of long-term risks;
- o maintaining a growing technological base in categories where government and industry must cooperate fully; and
- o promoting basic science and engineering research.

Federal Support for R&D

The Government supports R&D in many ways -- by sponsoring Federal laboratories; by providing over $40 billion annually for R&D; by contracting for R&D with research universities, colleges, and businesses; by allowing tax credits for industry-conducted R&D; and by encouraging entrepreneurs. The United States spent $4.6 billion for R&D in telecommunications in 1980 -- $2.8 billion in private sector investment, and $1.8 billion in Federal Government expenditures -- mostly through the Department of Defense (DOD).[6] DOD will increase funding for R&D in intelligence and communications[7] from $1.13 billion in 1981 to $2.77 billion in 1983.[8]

About one-half the total Federal obligations for all basic research are made to support researchers in universities and colleges, who account for about one-half of all basic research conducted in the U.S.[9] The key questions here are:

- o are we investing enough in the proper areas and through the most suitable institutions?
- o how can we measure and project U.S. progress in contrast with our most aggressive international competitors?
- o should we discount some DOD-funded research as a generally inadequate substitute for civil agency-funded research, in terms of contributing to the U.S. technological position?
- o where can improvements be made (in terms of removing obstacles to private sector investment, in transferring technology to the private sector, in providing incentives to private sector investment)?
- o what are the most promising opportunities for advancement of technologies, and are we positioning ourselves properly?

Federal Laboratories

Continued efforts to improve the role of Federal laboratories will ensure that appropriate types of R&D are undertaken, that long-term, high-risk programs with significant potential commercial pay-off are funded, that suitable industry collaboration is involved, and that research results are promptly transferred to the private sector. NASA's Advanced Communications Technology Satellite program is an example of a Federal laboratory targeting high-risk, long-term research based on industry advice, and contracting out a substantial portion of the research, thereby rapidly transferring research findings.

Planned budget reductions in the Space Shuttle program between now and 1994 -- and termination of NASA production of Atlas and Delta-Centaur rockets, the main launching vehicle for communications and other commercial satellites -- are prompting the private sector to consider increasing its investment in these activities.[10] Partly as a consequence of the shift in NASA involvement, delays in the Space Shuttle program, and the $1 billion investment of the European Space Agency for development of the Arianne rocket, strong foreign competition for the commercial space launch market is expected within a few years.[11]

Industry Support for R&D

Despite economic recession, industrial research and development spending surged in 1981. In particular, companies producing computer peripheral equipment and providing data processing services in the information processing industry led other industries with a 34.2 percent increase in R&D spending (to $344 million). Office equipment manufacturers in the information processing field increased R&D expenditures by 24.8 percent, and telecommunications firms' expenditures increased by 20.1 percent.[12]

The Semiconductor Industry Association has established a Semiconductor Research Cooperative which will channel $40 million into university laboratories by 1986.[13] Advances in telecommunications and information technology, as well as in applications, have continued to progress in areas ranging from large-scale computer and personal microcomputers (now a $6 billion market, with continued rapid growth projected), to subscriber loops, earth sensing from space, and office automation and medical electronics.[14] Foreign countries, such as Japan, West Germany, and the United Kingdom, however, are also very active in R&D in many of these areas -- especially in integrated circuits, in fiber optics and opto-electronic components, in portions of the satellite communications components field, and in office automation. While it is unclear whether there is a shortage of industry support for R&D, a comprehensive summary of R&D investments among the world's leading competitors -- including amounts, rate of change, areas of focus -- would be helpful in assessing the comparative U.S. position.

U.S. Patents and Licensing Trends

Trends in U.S. Patents. The number of U.S. patent grants (U.S. origin plus foreign origin) in telecommunications increased from an average of 174 in the 1969-70 period to 420 in 1981-82, a growth of over 140 percent. The proportion of foreign-origin patents to U.S.-origin patents has risen from 30 percent to 36 percent for these periods.[15] Japan's share of U.S. foreign-origin patents has risen to slightly more than 50 percent in the 1981-82 period from 34 percent for the 1969-70 period.[16]

During the 1978-80 period, the actual growth of U.S. patent awards in semiconductors and circuits improved by 48 percent, indicating substantial activity in solid state technology.[17] The General Purpose Programmable Digital Computer Systems and the Miscellaneous Digital Data Processing Systems subclasses commanded a major share of patent activity.

Trends in Foreign-Owned U.S. Patents. An area worthy of further investigation is the changing role of foreign multinational corporations (FMNCs) in the United States. A recent study by the U.S. Patent and Trademark Office[18] notes that foreign direct investment in the U.S. has been growing at an average rate of 20 percent since 1973. Part of the study focused on the largest FMNCs -- five European and five Japanese -- and sought to gain insights into questions including how much U.S. technology the FMNCs control, how much of their R&D is performed in this country, what technologies they patent, and what the trends are in their patent activities?

The study scope was not limited to particular technologies (although four Japanese and one European FMNCs are in the electronics and appliance industry group). It indicated that:

- one out of eight U.S. patents of foreign origin was owned or controlled by only 10 FMNCs during the time frame of the study, an indication of their strong position in the international technology marketplace.
- foreign-origin patents, as a whole, increased from 20 percent of all U.S. patents in 1963-66 to 40 percent in 1980.

- o ten FMNCs own or control, on average, 4.7 percent of all U.S. patents granted each year.

- o while the five European FMNCs had an average of about 10 percent of their patents filed as being of U.S.-origin[19], during 1976-80 the five Japanese FMNCs averaged about 0.4 percent.

- o the percentage growth[20] of the five Japanese FMNCs patents compared to the average[21] was especially high in the Computers and Data Processing Systems class (71.6 percent versus 49.8 percent) and in Static Information Storage and Retrieval (67.6 percent versus 47.1 percent), but slightly below the average for telecommunications (44.4 percent versus 47.2 percent).

Licensing of Technology to Overseas Companies. A recent survey of 161 companies' attitudes, based on prior experience with licensing technology to overseas companies,[22] indicates that the predominent result of the overseas licensing was long-term damage to the licensors. The consensus conclusion, based on comments of 107 of the companies, is that U.S. technology manufactured abroad by foreign firms finds its way into the U.S. domestic market.

Industry sources also note that Comsat -- and formerly AT&T -- are required to license their technology to other companies. Foreign firms that establish U.S. subsidiaries have made use of this regulation to acquire U.S. technology. Reciprocal licensing arrangements are not prevalent in foreign countries.

U.S. Restrictions on Technology Transfer

The central issue of technology transfer is balancing national security goals with other national goals, such as preserving First Amendment rights and the tradition of open academic exchange, while at the same time encouraging invention, commercialization, and exports.

National security interests sometimes require restrictions of technology transfer in the form of products, industrial processes, designs, and technical data that might aid current or potential foreign adversaries in a direct military way, or indirectly, by providing them with resources that otherwise would be allocated from military endeavors. The range of technologies that can be restricted includes telecommunications and information technology.[23] The mechanisms for effecting restrictions include: the Export Administration Act and the Inventions Secrecy Act, both administered by the Commerce Department; the International Traffic in Arms Regulations, administered by the State Department; and the national security classification system.

Studies have been ordered by Congress through the Department of Defense. Some of these studies recommended that restrictions on technology transfer affecting "cornerstone" technologies, processes, and designs be minimized and clarified. In the most recent effort toward clarification, the Commerce and Defense Departments were to prepare a joint Militarily Critical Technology List in 1982. There have been difficulties in implementing such recommendations, however, and some public confusion still exists.

The United States has the most stringent restrictions on technology transfer in the Western World. While there is little doubt that national security interests justify restrictions on transferring critical technology overseas, there is also a pressing need to streamline these restrictions as much as possible, while balancing them with overall U.S. policy goals. With clarified guidelines to protect national security goals, the private sector will be able to compete more effectively in the international marketplace through rapid technological advancement.

U.S. Entrepreneurs

Entrepreneurs play an important role in the United States in advancing technology, in providing employment opportunities, and in sustaining the competitive position of the United States. Collectively, American entrepreneurs represent a critical component of the nation's vitality, contribute to U.S. exports and have created approximately 70 percent of all new private sector jobs during the last decade. California's Silicon Valley alone adds over 25,000 new jobs annually, and Massachusetts' small and mid-sized high-tech firms have sparked an industrial renewal that now accounts for over 250,000 manufacturing jobs.

Associations of entrepreneurs (e.g., National Association of Small Business Investment Companies, American Electronics Association, American Business Conference), generally claim that the U.S. Government does not understand their needs, and favors big business in tax legislation. Entrepreneurs recognize "protectionism" policies as anathema to innovation and a process that leads to higher consumer prices. A recent GAO survey of 72 young "high-tech" companies started with venture capital shows that more than 15 percent of their total sales ($900 million) were to overseas customers.

A substantial amount of U.S. innovation is provided by small businesses and independent inventors. Federal support for innovative research by small businesses can be found in recent legislation, which requires Federal agencies to increase their R&D activities with small businesses, and which creates a new Small Business Administration information service to keep small businesses alert to opportunities for participation.[24]

Small, high-technology businesses are extremely sensitive to tax laws and policies, such as capital gains taxes, accelerated depreciation, R&D tax credits, and borrowing capabilities.[25] All of these affect the ability to generate investment capital to finance expansion. If tax policies encourage entrepreneurs, the long term benefits to the United States could be significant.

Foreign Strength in Key Technological Areas

Although the levels of U.S. R&D expenditures and growth in U.S. patents give the appearance of well-being, other factors suggest significant risks in some key areas. One such factor is the rising cost of continuing technological advancement, coupled with an uncertain pay-off in the distant future. An example of the problem of rising cost and questionable long term compensation is in the field of supercomputer technology, where there has been an erosion of the U.S. leadership position over the past 20 years.

Supercomputers. Many U.S. companies once were active in supercomputer development. The high expense and risks for this low volume market, however, have reduced their number to two -- Cray Research, Inc. and Control Data Corp.[26] As of June 1982, as few as 50 U.S. supercomputers were in operation worldwide -- 38 in the United States, 10 in Europe, and two in Japan. U.S. Government laboratories have 25, while a few private companies use the remainder, thus accounting for the absence of national support for supercomputer development.[27]

Japanese industry recognizes the importance of supercomputer research and has instituted an ambitious joint government-industry program to advance the state of the art in this and other selected technologies. One large program, the National Superspeed Computer Project, is aimed at developing a computer more powerful than any now available. This $200 million program is jointly funded by the government and six major Japanese companies, with 85 percent of the research contracted among the companies. Another large project is aimed at developing fifth generation computer technology with new capabilities in problem-solving, man-machine interfaces, and cognitive processes. A number of U.S. scientists who recently visited Japan and were briefed on these programs were generally confident of their success, while others expressed some doubt.[28]

A major recent report[29] on the U.S. position in supercomputers in relationship to the positions of West Germany, France, Great Britain, and especially Japan, concludes that the United States is rapidly losing its leadership. The report observes that the United States retreated from its support of large scale computing in universities and elsewhere during the 1970s when other countries' support grew substantially. An expert panel concluded that "there is little likelihood that the United States will lead in the development and application of this new generation of machines." The final report provides well considered recommendations for improving the U.S. position.

Integrated Circuits. During 1982, U.S. leadership in semiconductor or chip technology was shocked by Japanese companies' early market entry with a highly reliable 64K byte memory chip, the first time a foreign country posed a serious threat to U.S. leadership in this area. This was quickly followed by development of a prototype Japanese 256K byte chip. In the mid-1970s, the Japanese committed $350 million to a joint government-industry research project for very large scale integrated circuits. Japanese companies spent nearly 20 times as much during this program, which resulted in Japan's leadership in world markets for the 64K byte random access memory chip,[30] and perhaps for the next generation, the 256K byte chip as well.

These examples illustrate how concentrated efforts by a foreign country can drastically increase its technological leadership in a relatively short time. Loss of technological leadership has a number of serious consequences ranging from loss of access to the most advanced machinery for conducting research and forfeiture of technology spinoff opportunities, to inability to undertake new applications available through more sophisticated technology. In addition, there is the possibility that advanced U.S. scientific and military research could become dependent on access to supercomputers of foreign manufacture.

Other Nations' Support of R&D

The United States lags most industrialized nations in providing government incentives for invention and innovation, such as cost-sharing, salary grants, or interest free loans to strengthen the industrial technological base. Cost-sharing of high-risk efforts and more advantageous tax treatment is commonplace in other countries. For example, West Germany's accelerated depreciation for buildings and general purpose equipment devoted to R&D is tending to shift an increasing amount of U.S. R&D abroad.[31]

Canada's Industrial Research and Development Incentives Act, and the Program for the Advancement of Industrial Technology, provide tax free grants-in-aid up to 50 percent of R&D operating costs and 50 percent of the cost of new facilities, excluding land. In addition, Canada reimburses private firms for the salaries of their technical personnel assigned to government-approved research projects and allows all research costs, undiluted by grants, to be deducted from earnings for tax purposes.[32]

Japan provides even more incentives, including outright subsidies, cash grants repayable out of successful projects, and long-term, low-interest rate loans from development banks. Tax benefits include a 20 percent tax credit for increases in industrial research over expenditures during a base period (limited to a maximum of ten percent of the corporations' tax liability). Excluded from taxable income are 70 percent of royalties received from the export of technology.[33] In the computer area the Japanese government, acting jointly with leading Japanese computer firms, spent over $1 billion on computer R&D programs between 1972 and 1982 to catch up with the United States.[34]

Israel provides R&D grants that cover 50 percent of costs where the product to be developed is intended primarily for export.[35]

Current Activities to Improve the U.S. R&D Position, and Continuing Problems

Improvements in Federal Policies

There is widespread awareness that the U.S. position in the commercial marketplace and in technological leadership is increasingly threatened by strong, aggressive foreign competitors. To bolster the U.S. position, a variety of steps are being considered or have been taken, which include those based on Congressional or Administration initiatives, or on recommendations from trade associations and the academic and scientific communities. Few of these efforts focus primarily on telecommunications and information technologies, however, with some notable exceptions, such as NASA's space shuttle, DOD's Very High Speed Integrated Circuits program, and part of NSF's program to establish technical centers jointly supported by universities and industry, all of which are Federally supported programs.

There is a need for continued efforts to bolster the ability of the U.S. telecommunications and information industries to remain competitive technologically. Such efforts include:

- removal of unnecessary Federal disincentives to private sector research investment (in the form of antitrust laws and procurement regulations);

- o creation of new incentives (such as improved tax deductions for R&D expenses and promotion of joint research projects); and,
- o Congressional attention to improve the nation's education system (e.g., to upgrade science and mathematics education in elementary and high schools, to facilitate universities' ability to upgrade obsolete laboratory equipment and to retain competent faculty, as well as to retrain the work force).

Efforts already under way to improve the U.S. technological position include:

- o added incentives for R&D investments through the Economic Recovery Tax Act of 1981;
- o broader use of limited partnerships;
- o changes in patent and technology transfer policies; and,
- o better alignment of Federal laboratory programs with industry needs.[36]

Some areas identified for further review, which are discussed below include tax policies, Federal funding of research, and support for industry research consortia.

Tax Policies. A study performed by Data Resources, Inc. (DRI) for Texas Instruments, Inc., indicates that a 25 percent credit on R&D spending starting in 1966 would have added 0.2 percentage point per year to annual productivity during 1966-77, 0.3 percentage point per year in 1978-87, and 0.4 percentage point per year in 1988-97. DRI estimated that a 50 percent tax credit would have the effect of returning R&D expenditures to their (1964) peak of 2.15 percent of GNP within eight years, and that it would take about 35 years to achieve the same level of productivity improvement with a 25 percent R&D tax credit.[37]

Incentives for Investments in Research. Federal tax laws[38] provide a credit for investments for certain qualifying R&D of 25 percent of the maximum incremental increase over the base period investment. This may be an overly restrictive ceiling that discourages additional investment. Similarly, many types of R&D are not tax deductible, such as research and development for quality control and testing. Firms that are new or starting up receive no initial inducement to invest in research, while limited partnerships receive no deduction incentives at all.

A review of existing tax policies is warranted to determine whether a more favorable investment environment can be structured[39] which would be more sensitive to the needs and special characteristics of high-tech firms. These firms are different from others in a number of ways:

- o They must make substantial, continual investment in R&D, which in turn necessitates rapid replacement of existing plant and equipment. Therefore, favorable tax policies relating to the treatment of R&D expenditures and short-lived equipment are important.

- o These companies often have rapid growth in sales -- between 15 and 30 percent -- and retained earnings that are generally insufficient to meet capital needs. Typically, new equity investment must be obtained as the primary means of meeting these needs. Tax policies that favor debt financing thus may not be especially useful in promoting U.S. high technology firms.

- o They often must have significant international sales and operations. To stay ahead of foreign competition, they must be competitive in world markets, and therefore, are especially sensitive to U.S. tax policies relating to foreign operations and to U.S. exports.

Federal Funding of Research. Federally funded research will increasingly focus on longer-term, higher-risk, basic research, such as the NASA Advanced Communications Technology Satellite program, in which the private sector is unlikely to invest adequately. There is also increased Government funding emphasis on contracting through research universities, as well as encouraging joint industry-university cooperation. Over the years, the National Science Foundation (NSF) established seven centers for research which are eventually to become self-sufficient. The most recently established center is for research in communications and signal processing at North Carolina State University, which will include both basic and applied research leading to industrial products and services. The NSF grant is $650,000 over five years, with each participating private company contributing approximately $50,000. Nine companies are taking part in this cooperative effort. MIT (in polymer processing) and Rensselaer Polytechnic Institute (in computer graphics) are among other universities participating.[40]

Cooperation between Federal laboratories and industry is also being encouraged. The Department of Commerce has begun working closely with the Federal Laboratory Consortium to encourage technology transfer from the hundreds of U.S. Government laboratories to state and local governments and to industry. Small companies, independent inventors, and universities at the periphery of an industry, have also accounted for many of the major inventions in the twentieth century.[41] Under NSF's Small Business Innovation Research Program, approximately 52 percent of the grants have gone to firms with 10 or fewer employees, and several new firms have been started as a result of these grants. Annual funding is expected to reach $150 million in grants through NSF and other agencies.

Support for Industry Research Consortia. Some high technology industries faced with very high R&D costs and intensely subsidized foreign competition, have proposed joint R&D consortia that would be immune from some conventional antitrust restrictions. Recently the Justice Department approved establishment of the Microelectronics and Computer Technology Corporation, a consortium of several major computer firms. Similar consortia may also be accorded freedom from any threats of Government antitrust prosecution. The position the Justice Department has taken may encourage improved U.S. leadership in selected technologies, such as computers and microelectronics.[42]

Continuing Problems

Education-Related Problems. In order to maintain a viable position in technology and international trade, it is essential to ensure adequate talent is available to perform the research necessary to develop new concepts and new products. America's strength in these fields has long been a functon of its engineering and scientific human resources. In 1981, approximately 18,000 students graduated with Bachelor's, Master's, and Doctoral degrees in high technology fields in the United States. Since 1979, however, there has been a declining number of students in these important fields.[43] The problem that began with inadequate elementary and secondary school student preparation in science and mathematics in the Sixties and the Seventies is now being reflected in the shortage of college engineering graduates. With a shortage of U.S. engineers, U.S. companies currently confront an additional problem in the form of proposed restrictions that would prohibit foreign nationals from being hired immediately after graduating from U.S. universities.

There are thus three problems contributing to the present engineering conundrum:

(1) Students entering higher-level education in science and engineering too often are inadequately prepared. This problem begins at the primary and secondary education level, where there is a lack of adequate teaching proficiency in mathematics and science.

(2) Competent high school and university instructors are leaving teaching careers, often to double their salaries in industry, producing a 40 to 50 percent shortage of engineering faculty.

(3) Much of the laboratory equipment in the academic environment has become obsolete. The 97th Congress considered several bills to

> provide tax incentives to industry to provide more modern equipment to higher level institutions. The same opportunities are not yet available for secondary school education.

President Reagan spoke concerning these educational deficiencies at a May, 1982, National Academy of Science Conference. He noted that elementary and secondary school science and mathematics had deteriorated to the point that they threatened "to compromise the nation's future ability to develop and advance our traditional industrial base and to compete in the international marketplace." He called remedial action "long overdue," and invited private industry to do more to help schools.[44] Two recently created organizations, the National Commission on Excellence in Education under the Department of Education, and the National Science Board's Commission on Pre-college Education in Mathematics, Science and Technology are seeking remedies for these educational inadequacies,[45] and Congress may soon consider proposals to provide assistance to improve elementary, secondary, and post secondary education in mathematics and science.

Education at the secondary level provides the basis for those pursuing higher education needed for development and innovation of products instrumental in determining the advancement of U.S. society and its standing in international trade. Even more importantly, the quality of technical and scientific secondary education is reflected in the majority of the population entering the labor force. The success of this country will depend upon how well prepared our children are to face the information oriented world that is emerging due to massive technological advancements today. From the standpoint of the needs of research and development, with little additional training these students will be the technicians who create and apply R&D products.[46]

One effort to modernize primary and secondary education is through introduction of computers at this level. Proposals have been made to allow tax incentives for industry to contribute computer equipment to primary and secondary schools, as well as to provide grants to states for electronic and computer technician vocational education programs.[47]

In regard to the issue of foreign national graduates, immigration policies also are relevant as a number of qualified potential research employees educated at U.S. universities are not U.S. citizens.[48] The basic problem is, of course, not the retention of foreign students, but rectifying educational deficiencies that lead to this lack of American engineering talent. These educational deficiencies could cause profound, long-term damage to the U.S. technological and international trade positions, as well as to national security. Twenty-four measures were proposed during the 97th Congress to deal with these issues.[49] Continued intensive attention is warranted until suitable remedies are found.

Bell Telephone Laboratories. Bell Telephone Laboratories (BTL), AT&T's research arm, has produced a remarkable number of technological advances for more than 50 years. BTL is recognized as an undisputed world leader, at the forefront of technologies such as semiconductors, computers, and lasers.

With a 1982 budget of $2 billion, of which $156 million was spent for basic research, funding for research has been generous as well as consistent. BTL has almost 20,000 patents in effect, and currently receives new ones at the rate of one per day.[50] Under the 1956 Western Electric antitrust consent decree, AT&T was required to license these patents to other companies. Consequently, it has 400 licensing agreements with U.S. companies and nearly 200 more with overseas corporations today.[51] Some of these licenses have spawned entire new industries, perhaps years sooner than they might otherwise have been launched.

In August 1982, the Justice Department ended its 1974 antitrust suit against AT&T. A major reorganization of AT&T was required. The terms of the consent decree may affect future BTL research and development in a number of potentially significant ways.

First, under the new consent decree, AT&T will no longer be required to license its patents. Second, BTL is less likely to continue its traditional pattern of publishing hundreds of scientific and technical papers annually without concern for commercial value or patentability. Third, the new consent decree would also limit the operating companies and the Long Lines Department to funding specific, directed research through BTL. Previously, BTL basic research was funded by all parts of the Bell System through payment of a general license fee.[52] Pressures on the unregulated portions of AT&T to be competitive in the marketplace are likely to result in more product-oriented research by BTL at the expense of basic research.[53]

The AT&T settlement does permit the retention of BTL as a single entity. The long-term consequences of the divestiture on BTL's continued ability to generate technological advances, however, are not predictable.

Recommendations for Stimulating U.S. R&D

Our review of the U.S. international position in R&D in telecommunications and information technologies shows significant shortcomings, and obstacles affecting continued advances by U.S. industry. The main observation is that foreign competition for leadership in these technologies is exceptionally aggressive, and has grown rapidly over the past few decades. The single most important conclusion drawn is that continued U.S. success will be very much dependent both on Federal funding for R&D as well as on policies that create an environment conducive to continued vigorous private sector investment in R&D.

Option: Declare R&D A National Priority

Over the past few decades achievement in scientific and technological fields has varied as a priority in the United States. Administrations have at times actively sought scientific and technological advice from the private sector and at other times avoided it. Government funding of nondefense R&D has generally risen between 1953 and 1982, with some exceptions, but frequently at a rate below that of inflation. Between 1981 and 1983, total Government nondefense R&D will decrease by $1 billion to $17.3 billion, a 5.7 percent drop before adjusting for inflation.[54] At the same time, the Administration's overall economic recovery program has restored needed stability and predictability with the result that the private sector has stepped up its support of basic research.

If the U.S. becomes determined to make every feasible effort to maintain a position of technological leadership, then this effort should become a national priority. Maintaining technological leadership will require reinforcing current activities to remove investment disincentives, as well as providing additional incentives for the private sector. As noted earlier, investment incentives may be achieved by:

- partial relaxation of some actual or perceived antitrust constraints on joint research activities (now underway);
- revision of procurement regulations that deny contractors full rights to patents developed under Federal programs (already accomplished for nonprofit companies and for universities,[55] but not for larger companies);
- improved tax credits for R&D expenses; and
- promotion of joint research projects.

Option: A New Federal Policy on Funding of R&D

The need to "[C]reate the incentives and the supply of technological innovation" is applicable to the telecommunications and information technologies as well as to others.[56]

Given the long term interests of the nation, aggressive Federal funding of R&D in telecommunications and information technology may be warranted.

Ideally, funding for this research and development would be consistent, generally increasing at least as fast as the inflation rate, and sufficiently flexible to counter foreign activities.

Most agree that projects in which Federal resources are allocated to assist private industry and universities build long-range cooperative research capabilities should be encouraged and strengthened. A review should be made to determine whether sufficient Federal grants and contracts for R&D in telecommunications and information technologies are targeted toward small businesses, independent inventors, and universities. Ongoing U.S. Government activities that stimulate innovation include:

- o focusing Federal policy to support basic research in R&D;
- o improving technology transfers from Federal laboratories;
- o finding incentives for industry support of innovation; and
- o steps toward establishing joint university-industry centers of technology.

Demand for innovation, competition, and economic reward are three factors that stimulate innovation. Federal policies can positively influence the last two factors by establishing a suitable environment with tax advantages, minimal regulatory impediments, and direct Federal funding of R&D. Methods should also be sought, however, to encourage business and industry to transform research results into commercial products as soon as practical. It has been estimated, for example, that a considerable amount of research never leaves government laboratories. Efforts should be continued to use tax credits, improved Federal procurement regulations and patent policy, and consortiums with industry to motivate increased innovation. The long-term benefits to the nation should appear in the form of increased exports, employment, tax revenues, and decreased dependence on foreign technology.

Option: Create a Government Structure to Obtain Industry Advice

The Office of Science and Technology Policy is charged with advising the President on the nation's science and technology needs, and currently obtains private sector advice through the White House Science Council. Additionally, a White House Conference on Productivity was mandated through legislation at the end of the 97th Congress,[57] and the creation of a White House Council on Industrial Competitiveness is under consideration. Outside advice to the U.S. Government for Federal funding of R&D is strongly encouraged in order:

- o to identify the best emphasis for Federal funding;
- o to obtain expert views for decisionmaking concerning targeted versus generic research;
- o to provide guidance on the best balance of support for research conducted by universities and by private industry;
- o to improve commercial exploitation of government funded R&D; and,
- o to evaluate the merits of joint industry and joint government-industry research projects.

Option: Increase Use of R&D Limited Partnerships (RDLP)

U.S. industry could make far broader use of limited partnerships in research and development as a major means of funding technological advances. Because of the current trend toward lower interest rates, coupled with the effect of recently enacted reductions in Federal taxes, the volume of funds available for venture capital is growing substantially.

As a first step toward forming an RDLP, private businesses could identify individual R&D projects which, when commercialized, would make substantial contributions to U.S. competitiveness. The criteria for undertaking these projects should be that they:

(a) require more than simply the financial or technical competence of one firm acting alone, and

(b) have good prospects for being commercialized within a short period, such as four years.

Subsequent to project identification, market and technical data should be assembled to prepare a "commercial opportunity package," showing market segments, expected domestic and foreign demand, linkages into user industries, patent needs, etc. Private business (with Government encouragement, and technical and organization assistance, as needed) should develop a research and development limited partnership to fund the chosen venture, and to manage the R&D. The RDLP can provide tens of millions, even hundreds of millions of dollars of off-balance sheet funding without any obligation of repayment if the R&D is unsuccessful. Depending on the structure, the General Partner (the key player in any RDLP) can exercise management control. If structured properly, major antitrust problems can be avoided. The limited partners benefit from liberal tax shelter provisions, and the expectation of royalties which can be taxed at long term capital gain rates. A sponsoring company can retain control of proprietary rights. Equity participation is also possible.

The Department of Commerce's Office of Productivity, Technology, and Innovation can provide assistance in the RDLP area.

Recommendations:

Appropriate Level for R&D Funding. The appropriate amount of U.S. Government funding for R&D might be determined through indirect measures, such as a comparison of the U.S. position with that of other nations in terms of technical advances, patents, and innovative product development, plus amounts expended for R&D, both in dollars and as a percentage of GNP, etc.[58] A comprehensive analysis would be useful for developing periodic projections of U.S. R&D funding in comparison with that of major competitors over the long term.

U.S. Patent Position versus that of Foreign Multinational Corporations. The U.S. Patent and Trademark Office initiated a major statistical effort in the mid-1970s to develop a comprehensive data base with which to examine trends in U.S. patents both by technology and by country of origin. A recent study cited previously[59] notes that very few foreign multinational firms (FMNCs) control a significant proportion of all foreign origin patents (one out of eight, in fact, for the ten FMNCs for the time frame studied). These USPTO patent analysis studies should be encouraged and expanded -- along with earlier warning methods -- to develop insights into areas of potential foreign penetration into U.S. and world markets.

Licensing of Patents Rights. A study cited earlier notes a survey of 161 companies' attitudes based on experience with licensing technology to foreign firms. The survey was primarily negative in tone, reflecting the companies' view that the licensees too often become competitors for U.S. or foreign markets. Further study is suggested to expand these findings, to target similar studies on selected areas of high technology in the future, and to advise U.S. industry of the results.

Trade Practices. Many studies have been undertaken that concentrate on the ability of foreign competitors to subsidize sales, marketing, and R&D for new products. Their ability to "dump" (sell below cost) products on the U.S. market; and their practice of providing low-cost, long-term loans for procurement has also been reviewed. Nevertheless, we recommend that further study be initiated to determine how our major foreign competitors fund or subsidize R&D in telecommunications and information technologies. The results of this type of study should be useful for projecting the future U.S. technological position in regard to that of foreign competition and for planning U.S. strategies.

Bell Telephone Laboratories

In view of the singular importance associated with the BTL's historically large production of innovative technology, future BTL progress should be monitored for indications of decline, and, if needed, alternatives to restore overall U.S. innovations should be explored.

U.S. Restrictions on Technology Transfer

We recommend that restrictions on the export of telecommunications and information technology for national security purposes be

- limited to that which is essential,
- clarified for the public, and
- orchestrated in a way that substantially narrows the negative influence on invention and exports, and retains an atmosphere of open academic exchange.

Implementation of the above recommendations will assist the U.S. telecommunications and information industry to remain competitive and competent technologically in an increasingly aggressive world market. In addition, periodic assessments of the position of the U.S. industrial sector in relation to foreign competitors would serve to point out when further adjustments are needed.

NOTES TO CHAPTER NINE

[1]In 1981, over 40 percent of U.S. exports were represented by high technology products, according to U.N. Series D Trade Data from ITA in the Department of Commerce.

[2]The growth in employment in high technology and its supporting industries is approximately 50 percent higher than the growth in all U.S. business employment, according to U.N. Series D Trade Data from ITA in the Department of Commerce.

[3]At the Conference on U.S. Competitiveness at Harvard University, John W. Kendrick's calculations noted that a full 40 percent of U.S. growth in productivity over the last 50 years must be attributable to technological innovation. Research Management, July 1982, p. 40.

[4]1983 U.S. Industrial Outlook, U.S. Commerce Department.

[5]In his news conference of January 5, 1982, President Reagan recognized this shift in the nation's economy. He stated, "I think in this restructuring of our workforce

we're going to find that there are industries that have traditionally been at a certain level in our country, and we're going to find that we have switched to other lines of industry; to service types of industry rather than being a 'smokestack' industry-type. . . . We're going to have to prepare for them."

[6] Electronic Industry Association, "Electronic Market Data Book 1982," p. 121, Table 5-10, 1980 data.

[7] More detailed data are not available. It is likely, however, that much of the DOD funding includes non-telecommunications technology.

[8] American Association for the Advancement of Science, "Research and Development, AAAS Report VII, Federal Budget -- FY 1983 Impact and Challenge 4," p. 27, Table 2-2.

[9] Special Analysis K, Research and Development, The Budget of the U.S. Government, 1983, OMB, February 1982, pp. 2-3.

[10] IEEE Spectrum, January 1983.

[11] "We Mean Business," Forbes, January 1983, pp. 76-78

[12] These figures are based on data provided by companies surveyed by Business Week. See "R&D Scoreboard 1981," Business Week, 5 July 1982, p. 54. See also Research Management, September 1982, p. 2.

[13] Research Management, September 1982, p. 3.

[14] Institute of Electrical and Electronics Engineers (IEEE), Spectrum. Special Issue on Technology '83, January 1983.

[15] The data on patent grants are based on statistics for the first half of 1981 provided by the U.S. Patent and Trademark Office and adjusted for the full year. Foreign patents in communications equipment and electronics components increased from 16 percent in 1963 to 27.9 percent in 1973. Also, see "Patent Activity and International Competitiveness," Research Management, Vol. 21, #6, pp. 34-37, November 1978.

[16] Japan's patents as a proportion of U.S.-origin patents rose from 10 percent to 37 percent during the same periods.

[17] USPTO, Technology Assessment and Forecast (10th Report), November 1981, Section I, Active Technologies, p. 16.

[18] USPTO, Technology Assessment and Forecast (10th Report), November 1981, Section II, Patent Trends.

[19] An application is of U.S.-origin if the first inventor listed on the patent has a U.S. residence.

[20] Percentage growth is defined by the USPTO as the number of patents granted in the three-year period (1978-80) divided by the number of patents granted in the six-year period (1975-80) taken as a percentage.

[21] Average percent growth is calculated from all patents in the class.

[22] Forbes, 1 August 1982. Also see Research Management, November 1982, pp. 2-3.

[23] All exports are restricted except those which are specifically exempted.

[24] Small Business Research Act of 1981, Pub. L. No. 97-219 (22 July 1982).

[25] "Why Entrepreneurs Trust No Politician," Washington Post, 16 January 1983 p. Bl.

[26] J.E.E. (Japan) Progress on IC Technology and Full Scale Development of Overseas Local Production, July 1982, p. 88.

[27] "Japanese Supercomputer Technology," Science 218, 17 December 1982, pp. 1189-1193.

[28] Id. at pp. 1189-1193.

[29] U.S. Department of Defense in coordination with the National Science Foundation, Large Scale Computing in Science and Engineering, report prepared by a panel of experts, 26 December 1982. The report is available from NSF.

[30] "Japan's High Technology Challenge," News Week, 9 August 1982, pp. 48-54.

[31] "Technological Innovation - Key to Productivity," Research Management, pp. 33-41, July 1982.

[32] Ibid, pp. 33-41.

[33] Ibid, pp 33-41.

[34] U.S. Industrial Outlook, 1983, pp. 27-28.

[35] Research Management, pp. 33-41, July 1982.

[36] White House, Office of Science and Technology Policy in cooperation with the National Science Foundation, Annual Science and Technology Report to the Congress, Report No. NSF 82-9, 1981.

[37] Research Management, July 1982, pp. 33-41.

[38] As modified by the Economic Recovery Tax Act of 1981.

[39] "High Technology Industries and Tax Policy in the 1980's," National Journal, 1 January 1983, pp. 46-49. See also "Why Entrepreneurs Trust No Politician," Washington Post, 16 January 1983, pp. Bl-2.

[40] Research Management, p. 3, November 1982.

[41] "Technological Innovation - Key to Productivity," Research Management, July 1982, p. 34.

[42] Washington Post, 28 December 1982.

[43] Research Management, November 1982, reports a declining number of qualified science and engineering students, with peaks for Bachelor, Master, and Doctors

degrees occurring in 1973, 1974, and 1979 respectively.

[44]Speech to National Academy of Science Conference, May 1982.

[45]Research Management, Industrial Research at Universities, July 1982, p. 6.

[46]The significance of the educational problem is pointed out in a recent GPO publication (P-95 No. 76), which observed that the total number of Soviet scientists and engineers engaged in R&D during 1979 was 57 percent more than the number in the United States.

[47]U.S. Congressional Record, 98th Cong., 1st Sess., 6 January 1982, Vol. 129, pp. H 97, E 83.

[48]Intel, one of the leading U.S. semiconductor manufacturers recruits about 30 percent of its employees as foreign nationals.

[49]See Appendix C on legislation relating to R&D in telecommunications and information.

[50]U.S. Congress, Hearings Before the House Committee on Science and Technology, Subcommittee on Investigations and Oversight and Subcommittee on Science, Research and Technology on AT&T, 97th Congress, 2d Session, 8 December 1982 Statement of C. Kumar N. Patel.

[51]Business Week, 5 July 1982, pp. 46-52.

[52]Id. at pp. 46-52.

[53]Contrary to this analysis, an AT&T spokesman notes that the corporation is committed to sustaining BTL's basic research vitality, and does not anticipate any precipitous decline in BTL's focus or funding in the near term.

[54]"Science Advice Through the Years," National Journal, 25 September 1982, p. 1635.

[55]See Patent and Trademark Laws Amendments, 1980, Pub. L. No. 96-517. H.R. 4564 and S. 1657, introduced in the 97th Congress, were aimed at establishing and maintaining a uniform Federal policy for Federally sponsored R&D for all government contractors.

[56]"Technological Innovation - Key to Productivity, Research Management, July 1982, pp. 33-41.

[57]See Pub. L. No. 97-367 (October 1982), pertaining to The White House Conference on Productivity.

[58]Some of these measures are compiled periodically by the NSF, AAAS, EIA, Business Week, and other services.

[59]USPTO, Technology Assessment and Forecast (10th Report), November 1981, Section II, Patent Trends.

APPENDIX C

Complete Contents List to Long Range Goals in INTERNATIONAL TELECOMMUNICATIONS AND INFORMATION An Outline for United States Policy

PART II—INTERNATIONAL ACTIVITIES AND U.S. GOVERNMENT STRUCTURE

PART III—MAJOR COMPONENTS OF POLICY FORMULATION

APPENDIXES